Cryptands and Cryptates

Cryptands and Cryptates

Luigi Fabbrizzi

Università di Pavia, Italy

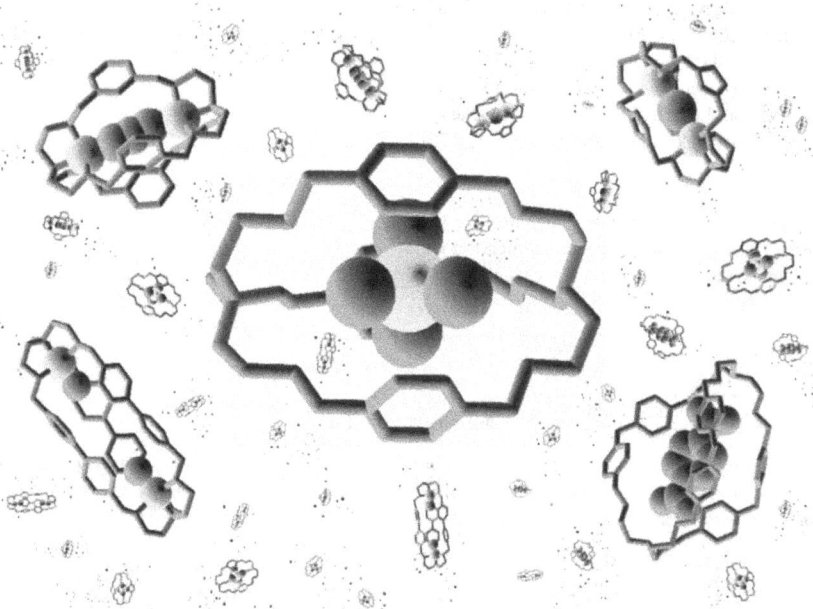

World Scientific

W JERSEY • LONDON • SINGAPORE • BEIJING • SHANGHAI • HONG KONG • TAIPEI • CHENNAI • TOKYO

Published by

World Scientific Publishing Europe Ltd.

57 Shelton Street, Covent Garden, London WC2H 9HE

Head office: 5 Toh Tuck Link, Singapore 596224

USA office: 27 Warren Street, Suite 401-402, Hackensack, NJ 07601

Library of Congress Cataloging-in-Publication Data

Names: Fabbrizzi, Luigi.

Title: Cryptands and cryptates / by Luigi Fabbrizzi (Università di Pavia, Italy).

Description: New Jersey : World Scientific, 2017.

Identifiers: LCCN 2017013393 | ISBN 9781786343697 (hc : alk. paper)

Subjects: LCSH: Macrocyclic compounds. | Supramolecular chemistry. |
    Ligands (Biochemistry)

Classification: LCC QD400.3 .F33 2017 | DDC 547/.5--dc23

LC record available at https://lccn.loc.gov/2017013393

British Library Cataloguing-in-Publication Data

A catalogue record for this book is available from the British Library.

Desk Editors: Kalpana Bharanikumar/Jennifer Brough

Typeset by Stallion Press
Email: enquiries@stallionpress.com

To my granddaughter Jennifer,
who did her best to prevent me
from writing this book
[*perhaps she was right*]

Foreword by Jean-Marie Lehn

While at the end of the road the path of research may look straight, seen from the beginning and along the way it can be rather erratic. The birth and development of the chemistry of cryptands and cryptates is no exception. The *Opus Magnum* that follows takes stock of the current state reached. The starting point may appear quite far off the line and indeed it was. The initial motivations and their translation to practice have been mentioned earlier (see *Angew. Chem. Int. Ed. Engl.* **1988**, *27*, 89–112). On the present special occasion, I may allow myself to briefly come back to the story of the origins.

In the course of the year 1966, while engaged in different topics of physical organic chemistry, my interest for the processes occurring in the nervous system led me to focus on the electrical events occurring in nerve cells, the action potential, which provides for the propagation of signals along the nerve axon and rests on changes in the flow of sodium and potassium ions across the nerve membrane. It was clear that there must be molecules in the membrane, probably proteins, capable of differentiating between sodium and potassium cations, two rather similar chemical entities with same charge and particularly close in size. Would there be small molecules that could also present such properties? Initial inspiration came from the fact that the cyclodepsipeptide antibiotic valinomycin had been found by C. Moore and B. C. Pressman in 1964 to mediate the transport of potassium cations in mitochondria. Other related cation-binding antibiotics were studied at the ETH in Zurich. It thus appeared that suitably designed synthetic cyclopeptides and analogues thereof could provide means of monitoring cation distribution and transport across membranes. On the other hand, earlier observations by H. C. Brown and G. Wilkinson had indicated that polyether-type compounds interact with alkali cations. Thus, when the cation-binding properties of macrocyclic polyethers (crown ethers) were reported by Charles Pedersen, these substances were perceived as polyether analogues of the cation-binding macrocylic antibiotics. Meanwhile, considering that in valinomycin the potassium cation was bound by inclusion within a cavity-like site, it had become clear that compounds containing a three-dimensional, spheroidal cavity, surrounding entirely the bound ion, would be better suited and should form stronger complexes than the rather flat macrocycles. Thus emerged the idea of designing macrobicyclic polyether ligands, which was also confirmed by the report in 1968 by C. H. Park and H. E. Simmons on the katapinates formed by encapsulation of chloride anions in macrobicyclic diammonium cations. The choice of a polyether structural type, much more stable towards reagents than the amide and ester functional groups of cyclic antibiotics, was in addition motivated by the perception of possible applications such as anion activation, since cation complexation within a large organic receptacle might represent a means of markedly increasing the reactivity of the counter-anion by hindering ion pairing, as was indeed shown later on to be the case.

Work was started in October 1967 with two graduate students, Bernard Dietrich and Jean-Pierre Sauvage, and the first macrobicyclic diazahexaoxa macrobicyclic ligand (termed [2.2.2]) was obtained in the fall of 1968. Its very strong binding of potassium ions was noted at once, and a structure where the cation was trapped inside the intramolecular cavity was assigned to the complex obtained, which was soon to be confirmed by a radiocrystallographic structure determination of the corresponding rubidium cation complex by Bernard Metz, Dino Moras and Raymond Weiss in 1970. To name this new class of chemical entities, I sought for a term, rooted in Greek and Latin that would also be equally suggestive in French, English, German, and possibly other languages: 'cryptates' appeared particularly well suited to denote a complex in which the cation was contained inside the molecular cavity, the crypt, of the ligand termed 'cryptand.' Other related macrobicyclic ligands were synthesised, a number of cryptates were obtained and their stability constants were measured. This work was also the occasion for my first foray into the world of patents. Indeed, Derek Barton, who was at that time consulting for Schering Co. in Bloomfield, New Jersey, recommended that we take a patent on these new class of compounds. This was done by Schering, but it delayed till 1969 our first two publications in *Tetrahedron Letters* in French! Three cryptands (designated as [2.2.2], [2.2.1] and [2.1.1]) were somewhat later made commercially available by Merck AG, Darmstadt, under the name Kryptofix, to be followed by others. Progressive generalisation of our initial work led to the notion of molecular recognition and later opened up a wider angle, including the domain of supramolecular chemistry. Such were the beginnings and the field took off, generating the imaginative and wide-ranging developments expertly presented in this book with great artistic talent by an important player in the game. All of us are most grateful to Luigi!

Speaking of its artistic resonances, chemistry is indeed in its essence a profoundly creative science, the art of matter, by its ability to create an endless set of novel chemical entities by re-composition of the bricks of matter, the elements, in an infinitely varied world of structures and functions. Their attractive plastic features are illustrated in many figures in the present volume. They may be inspirations for artists, as in the picture shown below, of the sculpture of a [2.2.2] cryptate by Béla Vizi of the University of Veszprém, a chemist turned artist.

Béla VÍZI (1936–) University of Veszprém

About the Author

Luigi Fabbrizzi was born in Florence in 1946 and obtained a degree in Chemistry from the University of Florence in 1969. From 1972 to 1980, he was a research assistant and a lecturer at the same university. In 1980 he moved to the University of Pavia, where he was a Professor of Chemistry until 2014, and is currently a Professor Emeritus. Since 2009, he has been a Honorary Professor at the East China University of Science and Technology of Shanghai. In 2010, he received the *Izatt-Christensen Award in Macrocyclic Chemistry*. His research interests cover several aspects of supramolecular chemistry including metal template reactions, molecular switches of fluorescence, anion recognition and sensing, and molecular machines based on ion translocation.

Contents

1

The Beginning of the Story: Crown Ethers

The birth of supramolecular chemistry, the science of non-covalent interactions, is conventionally associated with the publication of a paper in 1967 by Charles J. Pedersen, a senior researcher at DuPont, Wilmington, Delaware.[1,2] Pedersen, who was 63 at that time, had investigated at DuPont for many years the autoxidation of petroleum products and rubber, a complicated process catalysed by trace metals.[3] Hence, he was interested to develop multidentate ligands aimed to suppress the catalytic activity of transition metal ions through complexation. In 1960, he started a project on the for design of ligands suitable for binding (and 'deactivation') of the vanadyl ion (VO^{2+}, oxovanadium(IV)) and focused his attention on phenolate derivatives. In particular, he considered that bis[2-(o-hydroxyphenoxy)ethyl] ether **3** in Figure 1.1, when deprotonated, could effectively sequestrate the oxocation.

The synthetic design of **3** involved protection of one phenolic group of catechol, **1** with 3,4-dihydro-2*H*-pyran, followed by reaction with the dichloro derivative **2** and deprotection with acid to give the desired product (route (i) in Figure 1.1). At the end, a poorly attractive creamy substance was obtained, which on purification gave a small quantity (0.4% yield) of

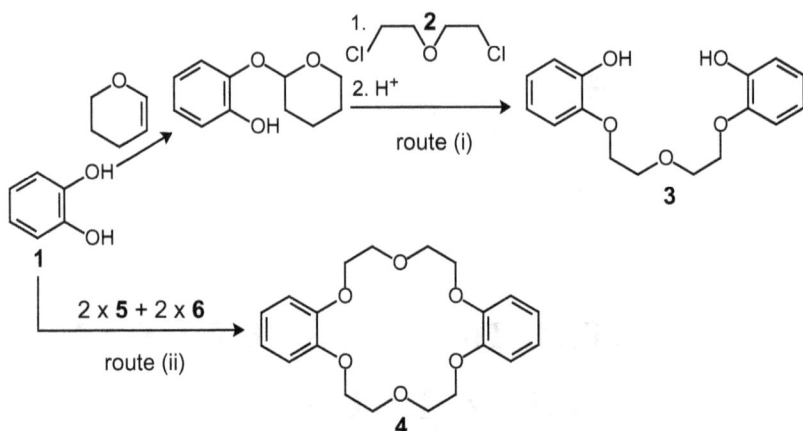

Figure 1.1.   Route (i): the aimed synthesis of bis[2-(*o*-hydroxyphenoxy)ethyl] ether (**3**), which, on deprotonation of the two phenolate groups, should bind the VO^{2+}; Route (ii): the 'serendipitous' synthesis of dibenzo-18-crown-6, **4**.

white fibrous crystals, which could not be the target compound. Most likely, a researcher in Academy, whether a student or a professor, eager to get (and to publish) results, would have left the crystalline by-product in a desiccator, in order to look for a different synthetic route to **3**. Pedersen, thanks to his 'natural curiosity' (and the absence of any academic pressure), decided to persist and 'study the unknown'. He demonstrated that the fibrous product was the cyclic polyether **4**, which had formed through 2 + 2 cyclisation of **2** and catechol **1**, which, used in excess, had remained as an impurity (route (ii) in Figure 1.1). In the following seven years, Pedersen, assisted by a skilful technician, Ted Malinowski, synthesised about 60 cyclic polyethers. He was fascinated by the structural formulae of these new compounds (and who wouldn't be?) when drawn in 2D (see for example formula **5** in Figure 1.2) and, even more, when drawn in 3D, taking into account the sp^3 hybridisation of carbon and oxygen atoms.

Imaginative 3D drawings like **5a** in Figure 1.2 motivated Pedersen to name this class of molecules 'crowns', thus avoiding the complicated IUPAC nomenclature (**5**: 1,4,7,10-hexaoxacyclooctadecane becomes 18-crown-6, where 18 indicates the number of atoms – C and O – constituting the ring and 6 the number of ethereal oxygen atoms). A lateral view of the

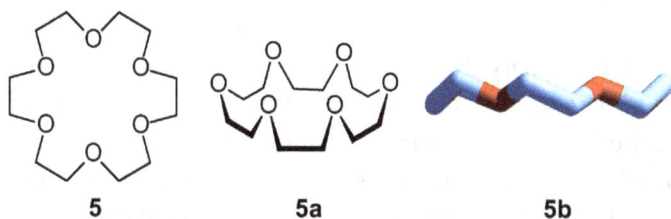

5   **5a**   **5b**

Figure 1.2. (**5**) Structural formulae of 18-crown-6; (**5a**) 3D structural formula, drawn to evoke a royal crown; (**5b**), crystal structure of [18-crown-6]·$(CH_3NO_3)_2$, according to a tube representation, lateral view (hydrogen atoms and solvate molecules have been omitted for clarity).[4] Most crystal structures of 18-crown-6 have a similar structure with all six oxygens convergent towards the inside of the macrocycle (*endo*). The first reported crystal structure refers to a molecule with two oxygens pointing to outside (*exo*)[5], thus not favourably oriented for coordination.

'real' molecule, from single crystal X-ray diffraction studies (**5b** in Figure 1.2),[4] showed a puckered arrangement of carbon and oxygen atoms, legitimising the crown name given.

The most striking property displayed by crown-ethers was their ability to bind and 'trap the otherwise recalcitrant alkali and alkaline-earth metal ions'.[3] Coordination chemistry of *d* block and *p* block metal ions was at that time a well-settled discipline, which had produced hundreds of multidentate ligands of varying sophistication and had established some basic principles and concepts. Pedersen unlocked the unknown realm of *s* block metal coordination chemistry, which, although not so rich and varied as *d* metals coordination chemistry, immediately displayed exciting applicative features in the field of organic chemistry, such as the capability of crown-ethers to solubilise insoluble inorganic salts in apolar and poorly polar solvents: the most spectacular example was probably the so-called 'purple benzene', referring to the solubilisation of $KMnO_4$ in benzene promoted by 18-crown-6 (**5**) or by dibenzo-18-crown-6 (**4**).[6] Dissolution of potassium permanganate (as the ion pair $[K(18\text{-crown-}6)]^+MnO_4^-$) allowed the oxidation of organic substrates in their solvents of choice by one of the most powerful oxidising agents of inorganic chemistry. In addition, the oxidising power of MnO_4^- was significantly enhanced by the fact that the anion in benzene and other apolar media is not solvated, but 'naked' and, consequently, much more active than in an aqueous solution.[7]

Moreover, the study of the interaction in a solution of cyclic poly-ethers of different cavity sizes with a homogeneous class of spherical cations of differing radii, e.g. alkali metals, offered the opportunity to evaluate geometrical effects on the thermodynamic stability of crown complexes and to introduce the new concept of geometrical selectivity. This is illustrated, for instance, by the diagram in Figure 1.3, in which log K values of the complexation equilibria of 18-crown-6 in MeOH, at 25°C, have been plotted against the ionic radius of the alkali metal ion.[8]

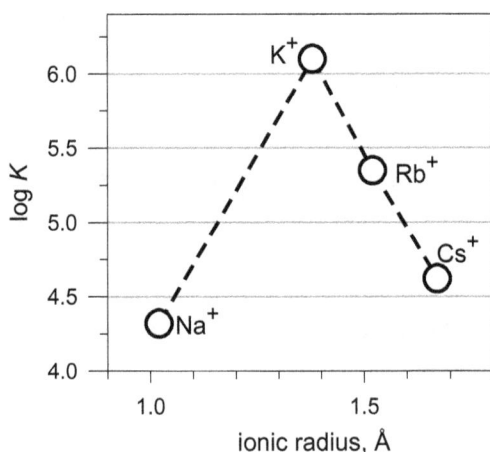

Figure 1.3. log K values for the equilibrium: $M^+ + L \leftrightarrows [ML]^+$ in MeOH, at 25°C. M = alkali metal, L = 18-crown-6 (**5**).[8]

A sharp peak selectivity was observed in favour of K^+, which apparently possesses the right radius to fit the cavity of the crown ether relaxed to its more stable conformation, thus allowing the formation of the strongest coordinative interactions (essentially electrostatic in nature). As a simplistic explanation for the peak selectivity behaviour, one could suggest that each metal ion lies in the plane of the six ethereal oxygen atoms and that the crown framework has to rearrange itself in order to guarantee the formation of a complex with the right metal–oxygen distances, either contracting (with Na^+) or expanding (with Rb^+ and Cs^+) its cavity. On these bases, the pattern of log K values should reflect the conformational energy cost associated with such a rearrangement. Things are not exactly so, as demonstrated by the X-ray diffraction studies. In fact, the crown framework maintains its conformation, while the metal ion moves up from the O_6 plane in order to reach the correct M–O distances. In particular, Figure 1.4 shows the position of each alkali metal ion with respect to the mean plane of six oxygen atoms of 18-crown-6, which have been linked together by red segments.

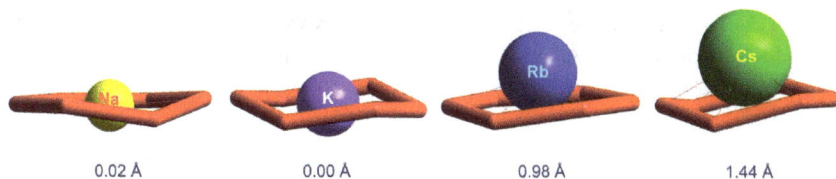

| 0.02 Å | 0.00 Å | 0.98 Å | 1.44 Å |

Figure 1.4. Position of alkali metal ions with respect to the mean plane of the six oxygen atoms of 18-crown-6, **5** (L), which have been linked by segments, to give a puckered hexagon. Values in Å give the distance between the metal and the mean least-squares plane of the six oxygen atoms of the crown ether. Crystal structures: [Na(L)][HfCl₅(THF)],[9] [K(L)] NCS,[10] [Rb(L)][GaCl₄],[11] [Cs(L)]NCS.[12]

The distance of the ion from the least-squares mean plane of the six oxygen atoms is indicated in Figure 1.4. K^+ is perfectly coplanar with the O_6 mean plane (distance 0.00 Å), and Na^+ is almost coplanar (0.02 Å). By contrast, Rb^+ and Cs^+ stay well above the plane (at 0.98 and 1.44 Å, respectively).

Let us first compare the behaviours of Na^+ and K^+, both well inserted inside the crown and apparently in the position to best profit from the coordination of the six oxygen atoms. It is useful to look at the metal–oxygen distances reported in Figure 1.5. For comparative purposes, Figure 1.5(a) shows the distances calculated between the oxygen atoms and the centroid of the metal-free crown ether (average distance: 2.86 ± 0.02 Å), which should be considered the ideal M–O distances for the strain-free coordination of a metal ion M^+. The K–O distances (average distance: 2.80 ± 0.03 Å) are very close to the ideal one and also match the 'natural' K–O distance. The natural distance is the one observed in an unconstrained situation, such as in the 1:3

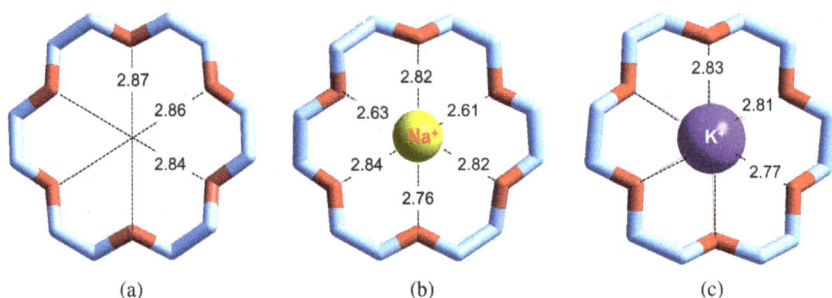

Figure 1.5. Top view of the crystal structures of 18-crown-6 and of its complexes with Na+ and K+; (a) distances in Å between ethereal oxygen atoms and the centroid of the six oxygens; (b) Na–O distances (Å); (c) K–O distances (Å).

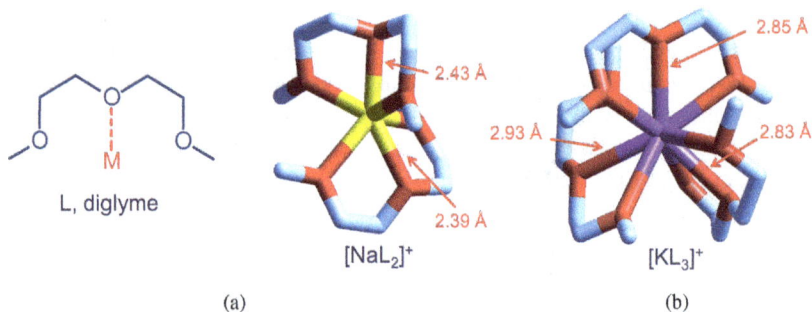

Figure 1.6. The crystal structures of the sodium and potassium complexes of diglyme (L): (a) [Na(L)$_2$]+,[13] and (b) [K(L)$_3$]+.[14] Bond lengths indicated by an arrow are considered the ideal bond distances for the envisaged metal ion.

complex of potassium with the open-chain ligand diglyme, as shown in Figure 1.6.

In particular, the ideal K–O bond length is considered as the distance between K+ and the mid-oxygen of the coordinated diglyme in the [K(L)$_3$]+ complex (see Figure 1.6(b)): 2.87 ± 0.06 Å. Thus, K+, when encircled by 18-crown-6, enjoys an especially favourable situation: (i) it forms K–O bonds of a length very close to that expected by the crown ether, which is therefore not forced to any sterical rearrangement; (ii) these K–O bonds are its own normal bonds, as far as distance, and presumably strength, are concerned. This explains the high stability in the solution of the [K(18-crown-6)]+ complex.

For sodium, properties are different. Figure 1.5(b) shows that four Na–O distances (2.81 ± 0.05) match quite well with the value required by 18-crown-6, but two do not (2.61 and 2.63 Å). In any case, Na–O distances in the crown complex are remarkably higher than the ideal distance as observed in the [Na(diglyme)]$^+$ complex (2.41 ± 0.02). This means that Na$^+$ stays in the plane of the six ethereal oxygens of 18-crown-6, but the macrocycle refuses to undergo any rearrangement to put its oxygen atoms at the required distances. As a consequence, Na$^+$ is lost in a vast space and establishes less intense interactions with ethereal oxygen atoms.

On the other hand, Rb$^+$ and Cs$^+$ are too big to reach coplanarity with the O6 donor set and place themselves above the plane at the appropriate distance to form bonds of natural length (see Figure 1.4): the higher the ionic radius, the more pronounced the displacement from the plane and, presumably the weaker the metal–ligand interaction.

Classical d block coordination chemistry originated and developed with the design of complexes of a unidentate ligand (ammonia), whereas s block coordination chemistry was discovered several decades later when cyclic multidentate ligands were designed. Metal ions with partly filled d orbitals benefit from Ligand Field Stabilisation Energy, a feature precluded to cations possessing an inert gas electronic configuration. Moreover, thanks to the poor shielding exerted by d orbitals, transition metals exhibit a comparatively high effective nuclear charge, which reinforces the electrostatic (or ion-dipole) contribution to the coordinative interactions.

Alkali and alkaline-earth metal ions cannot take advantage of the above features. They are hydrated in aqueous solution, but, on precipitation as salts, they do not keep molecules from the hydration sphere. Moreover, if the isolated salt contains crystallisation water molecules, they are not bound to the metal, but rather to the counteranion(s) through hydrogen bonding, a behaviour which demonstrates the intrinsically weak metal–donor atom interaction. This is not the case of transition metal salts, which often crystallise as hydrates, with water molecules directly bound to the metal ion, according to a defined geometrical arrangement (in most cases octahedral). The ethereal oxygen atom is a stronger donor than the oxygen atom of water, but no alkali and alkaline-earth complexes with unidentate ethers have ever been isolated. The *chelate effect*, which dominates transition metal chemistry, does not operate in s block coordination chemistry, and only poorly stable complexes with open-chain polyethers have been observed. Pedersen demonstrated that *s* block metal ions can give stable complexes with polyether ligands only if profiting from a *macrocyclic effect*, whose nature is illustrated in Figure 1.7.

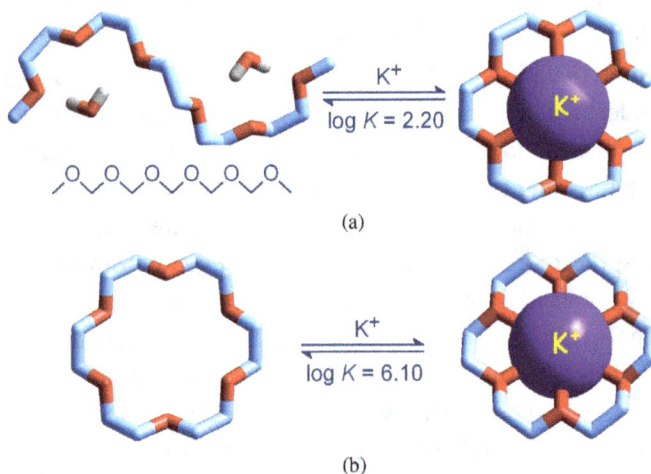

(a)

(b)

Figure 1.7. (a) The crystal structures of pentaglyme dihydrate (the open-chain analogue of 18-crown-6)[15] and its potassium complex salt [K(pentaglyme)](CF$_3$SO$_3$)[16] (hydrogen atoms of the polyether and counter-anion have been omitted for clarity). Pentaglyme crystallises with two water molecules, each bound to three ethereal oxygen atoms through hydrogen bonding; (b) the crystal structures of 18-crown-6 (**5**) and its potassium complex salt. Log K values determined in MeOH at 25°C.[8]

The open-chain sexidentate polyether pentaglyme (pentaethyleneglycol-dimethylether whose crystal and molecular structure is shown in Figure 1.7(a)) wraps the K$^+$ ion for complexation to give a 1:1 complex. In doing so, the chelating agent spends both entropic (associated with the loss of freedom) and enthalpic (generated by intra-ligand steric repulsions) energy. Thus, the intrinsically low energy contribution from metal-donor atom interactions is substantially cut down by the energy paid for cyclisation: the log K value of the complexation equilibrium (a) of Figure 1.7 is 2.20 (in MeOH at 25°C). On the contrary, no energy cost must be paid on complexation of K$^+$ by the crown, which is favourably prearranged for encircling the cation: the molecular structures in Figure 1.7(b) show that metal complexation does not require any conformational rearrangement of the crown. The energy released up on the formation of the coordinative bonds is still small, but it is not counterbalanced by any term related to ring closure and allows the formation of a stable complex in solution: log K for equilibrium (b) in Figure 1.7 is 6.10 log units, i.e. nearly four orders of magnitude higher than for the non-cyclic counterpart.

The higher stability of metal complexes with cyclic ligands compared to open-chain analogues containing the same type and number of donor atoms is ascribed to the macrocyclic effect (ME), which is quantitatively expressed by the ratio of the complexation constants of cyclic and open-chain complexes. In the present case, ME = 7.9×10^3. The macrocyclic effect has been formulated for $3d$ metal complexes with tetramines (and usually corresponds to 4–5 orders of magnitude), but it has a general relevance: cyclic ligands will form more stable complexes than their non-cyclic analogues because they do not have to spend any cyclisation energy. Such an energy cost has already been paid over the course of their synthesis. Moreover, a further significant role is played by solvation: the open-chain polyether can expose its oxygen atoms to the solvent molecules with which it can comfortably interact. In this way, the pentaglyme ligand shown in Figure 1.7(a) crystallises with two water molecules, each one donating two hydrogen bonds to two ethereal oxygen atoms. On the other hand, the ethereal oxygen atoms of the crown converge towards the cavity and can hardly be accessed by solvent molecules. Thus, the macrocycle in metal complexation does not suffer from the endothermic desolvation process experienced by its open-chain counterpart — a further contribution to the macrocyclic effect.

2

The Birth of Cryptands and Cryptates

In the same year Pedersen published the papers on crown ethers, Jean-Marie Lehn, a young Professor at the Université Louis Pasteur, Strasbourg, was studying the chemical processes that occur in the nervous system. He was aware that bioelectrical stimuli originate from changes in the concentration of Na^+ and K^+ inside and outside the nerve cell and that the natural poly-oxa macrocycle valinomycin (**6**) mediates their transport across the membrane. Valinomycin shows an appealing circular formula, containing 6 ethereal oxygen atoms and 12 carbonyl oxygen atoms as potential donors. Valinomycin, in crystallographic reality (Figure 2.1(a)),[17] shows a less fascinating, poorly symmetric structural organisation. However, on interaction with K^+, it rearranges to provide a nice spheroidal cavity and establishes with the included cation six coordinative interactions using metal-carbonyl oxygen atoms as donors, according to an almost regular octahedral geometry (see crystal structure in Figure 2.1(b)).[18]

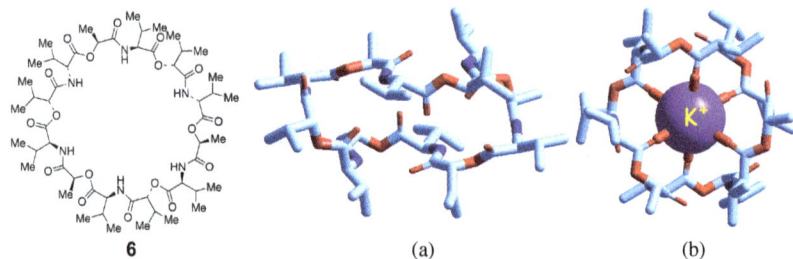

Figure 2.1. The crystal and molecular structure of valinomycin (**6**; a)[17] and its potassium complex (b).[18] Hydrogen ions and the picrate counteranion of the complex have been omitted for clarity.

Based on these observations, Lehn considered that 18-crown-6, which offers a planar cavity, was not the best artificial substitute of valinomycin and devised the design of a receptor able to provide a 3D spheroidal cavity. In this sense, Lehn and his students Bernard Dietrich and Jean-Pierre Sauvage accomplished the synthesis of the macrobicyclic ligand **8**, as illustrated by the scheme in Figure 2.2.[19]

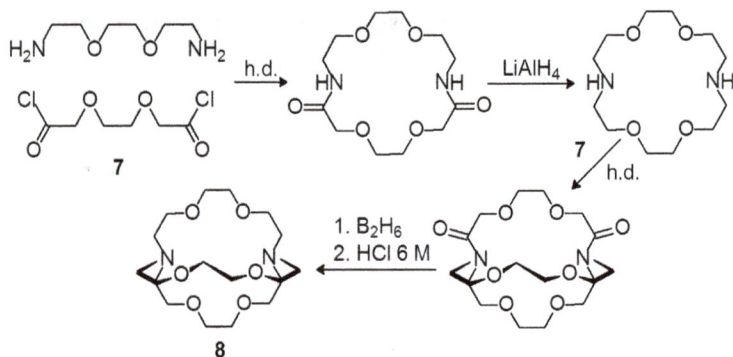

Figure 2.2. The synthesis of cryptand **8**, involving two high dilution (h.d.) cyclisation steps.[19]

In particular, molecule **8** contains two tertiary amine bridgeheads linked together by three $-(CH_2)_2O(CH_2)_2O(CH_2)_2-$ chains and is expected to provide a spherical cavity, with six coordinating ethereal oxygen atoms, strategically positioned in the three dimensions, possibly at the corners of

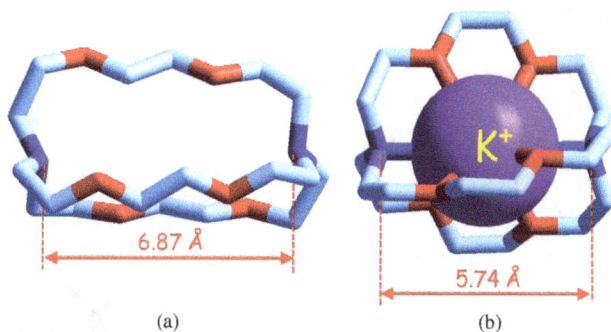

Figure 2.3. The crystal structures of cryptand **8** (L)[20] and its complex salt [K(L)]I[21] hydrogen atoms and iodide counterion have been omitted for clarity. Upon metal compl-exation, the ellipsoidal receptor rearranges to a conformation providing a spheroidal cavity with a significant reduction of the N_{tert}---N_{tert} distance.

a trigonal prism. However, X-ray diffraction studies (Figure 2.3(a)) showed an elongated arrangement of the receptor, providing an ellipsoidal cavity from which the six ethereal oxygen atoms diverge.[20]

On reaction with a potassium salt, **8** gives a 1:1 complex, whose structure is shown in Figure 2.3(b).[21] Indeed, on complexation, receptor **8** undergoes a conformational rearrangement to offer the metal a spheroidal cavity, in which the spherical cation fits well. Moreover, K[+] interacts not only with the six oxygen atoms but also with the two bridgehead tertiary amine groups.

The IUPAC name of **8** (1,10-diaza-4,7,13,16,21,24-hexaoxabicy-clo[8.8.8]hexacosane) was too complicated, and so the authors looked for a simple and possibly expressive nickname. If Pedersen took inspi-ration from the *shape* of receptor, Lehn looked at the *function* and named **8** *cryptand*. The term results from the combination of the Ancient Greek adjective 'kryptos' (κρυπτός, hidden, secret) and 'ligand': the ligand that keeps metal ions hidden. A different imagina-tive interpretation would refer to *crypt*, the ritual chamber located beneath the floor of Medieval churches, in which sacred items and rel-ics of Saints were stored. One could guess that Lehn, in coining this term, was inspired by the beautiful Crypt of the Strasbourg Cathedral, shown in Figure 2.4.

Figure 2.4. Cathedral of Our Lady of Strasbourg: the Roman crypt (XI century).

In any case, the word 'cryptand', in view of its expressiveness and also for the intrinsic appeal of the molecule, started the trend of a unique chemical nomenclature and, since then, a variety of *X-and* terms have been coined, in which the *X*-root, in most cases of Greek or Latin origin, recalls either the shape or the function of the molecule, e.g. spherands, carcerands, torands, scorpionands, catenands, borromeands, etc. Following such a pervasive fashion, there was also an attempt (not very successful) to rename crown-ethers *coronands*.

If cryptand is the 'ligand' (or receptor), the associated complex is the 'cryptate'. It now has to be verified whether cryptates are more stable than the corresponding complexes of crown-ethers and of the natural counterpart valinomycin.

In Figure 2.5, log K values for the complexation equilibria $K^+ + L \leftrightarrows$ $[K(L)]^+$ in MeOH and water, at 25°C, are reported for comparison (L = 18-crown-6 (**5**), cryptand (**8**), valinomycin (**6**)). The cryptand forms the most stable complex in both MeOH and water. In water, where a quantitative comparison can be made, the cryptate is 5000-fold more stable than the crown complex. However, the comparison is not totally correct

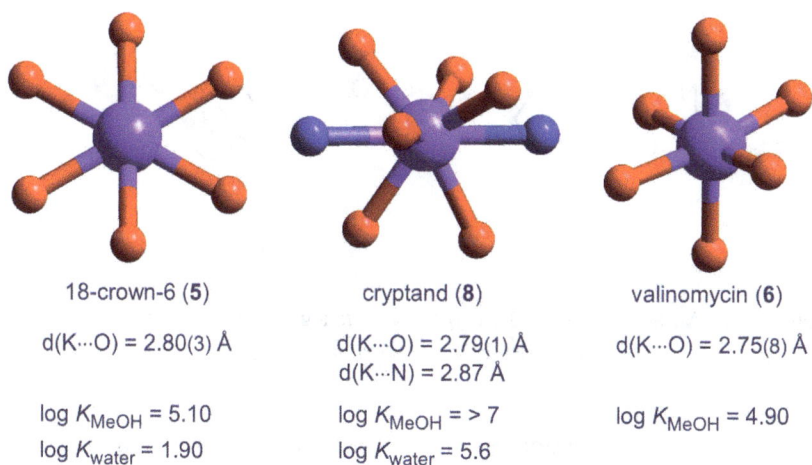

18-crown-6 (**5**)	cryptand (**8**)	valinomycin (**6**)
d(K⋅⋅⋅O) = 2.80(3) Å	d(K⋅⋅⋅O) = 2.79(1) Å	d(K⋅⋅⋅O) = 2.75(8) Å
	d(K⋅⋅⋅N) = 2.87 Å	
log K_{MeOH} = 5.10	log K_{MeOH} = > 7	log K_{MeOH} = 4.90
log K_{water} = 1.90	log K_{water} = 5.6	

Figure 2.5. Coordination geometries and average bond distances of K^+ complexes of 18-crown-6, 2.2.2-crypt and valinomycin (in parentheses uncertainties in the last figure). log K values of complexation equilibria determined in MeOH and water at 25°C.[22]

because the cryptand establishes with the metal two additional interactions with the two bridgehead nitrogen atoms, which are not there to play a mere architectural role. On the other hand, a homogeneous comparison can be made by comparing log K values of 18-crown-6 and valinomycin in MeOH, both coordinating K^+ with six oxygen donor atoms. Paradoxically, the valinomycin complex, showing an almost regular octahedral geometry, is less stable (only by 0.2 log units) than the crown complex, in which donor atoms are placed at the vertices of a hexagon. The hexagonal coordinative arrangement is more crowded and intrinsically less favourable than the octahedral one. However, the higher stability of the [K(18-crown-6)]$^+$ complex seems reasonably to derive from the nearly perfect arrangement of the uncomplexed crown, which, on incorporating the metal, has to pay no conformational energy cost (as shown in Figure 1.7(b)). On the other hand, Figure 2.1 shows that valinomycin, on complexing K^+, undergoes a significant endergonic rearrangement, which cancels the advantages of the 3D coordination.

Chapter 1 has shown that a significant macrocyclic effect exists in complexation of polyethers. Now, we look at whether a cryptate effect exists and how it can be quantitatively assessed.

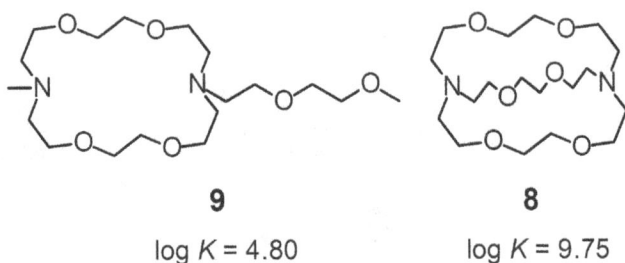

9

log K = 4.80

8

log K = 9.75

Figure 2.6. An assessment of the *cryptate effect*: log K values for the complexation equilibria $K^+ + L \leftrightarrows [K(L)]^+$ in MeOH/H$_2$O 95/5 v/v at 25°C (L = **9**, **8**).[22]

The correct evaluation of the cryptate effect has been made by comparing log K values for the potassium complexation equilibria involving cryptand **8** and the functionalised macrocycle **9**, an 18-membered N$_2$O$_4$ crown, bearing an *N*-linked (CH$_2$)$_2$O(CH$_2$)$_2$OCH$_3$ dangling chain (see corresponding formulae in Figure 2.6). The cryptate complex is nearly 10^5-fold more stable than the crown complex, presumably because it does not have to spend enthalpic and entropic energies to immobilise the side-chain as **9** does. In this sense, a cryptate effect exists and has a magnitude even higher than that of the macrocyclic effect.

3

Recognition of the Radius of *s*-Block Metal Ions by Cryptands

Recognition, which refers to the selective interaction of a receptor with a substrate, is one of the most investigated fields of supramolecular chemistry. The substrate is a *convex* chemical entity whose recognition is sought, the receptor is a *concave* molecular system designed for establishing favourable interactions with the substrate. The selective interaction of the receptor with a substrate, in the presence of other competing substrates, is called recognition. Figure 3.1 pictorially illustrates the principle of recognition, based on shape/size matching of the receptor's cavity and substrate.

A basic prerequisite for selectivity is that the receptor–substrate interaction is fast and reversible, so that, in the presence of a variety of substrates of varying size and shape, the receptor, through a *trial-and-error* mechanism, can eventually find its kindred spirit.

In coordination chemistry, the substrates are the metal ions, while receptors are called *ligands*. Transition metal ions of a given electrical charge have similar ionic radii, and discrimination of substrates is solely dominated by Ligand Fields effects. For instance, in the case of divalent 3d metal ions, the stability sequence rigidly follows the Irving–Williams

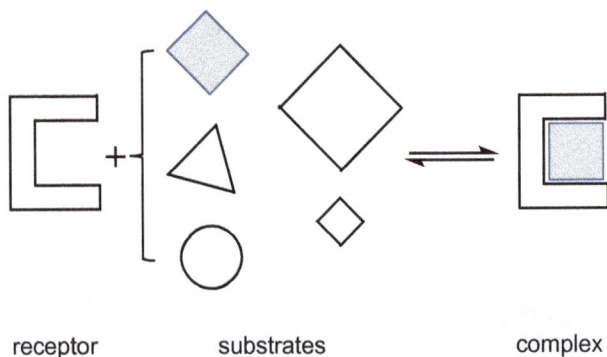

receptor   substrates   complex

Figure 3.1.   The principle of shape/size matching in the selective interaction of a receptor with its partner among a variety of substrates.

series ($Mn^{II} < Fe^{II} < Co^{II} < Ni^{II} < Cu^{II} > Zn^{II}$).[23] Thus, the design of a receptor capable of selectively interacting with, say, Mn^{II} in the presence of Cu^{II} can hardly be realised.

On the other hand, *s*-block cations possess a closed-shell electronic configuration, are spherical and differ from each other mainly by the ionic radius. Thus, the principle of shape/size selectivity should operate properly, which suggests that (i) receptors for alkali and alkaline-earth metal ions should provide a spheroidal concavity and (ii) selectivity should simply result from the more or less favourable matching of the size of the cation and that of the cavity. In this sense, cryptands are expected to do an excellent job.

Figure 3.2 shows a diagram in which log *K* values for the complexation of K^+ by cryptand **8** (circles, MeOH/water, 95:5 v/v, 25°C) have been plotted against metal ionic radius. The diagram displays a well-defined peak selectivity in favour of K^+, which has the right radius to fit the spheroidal cavity provided by cryptand **8**. Such a selectivity effect is distinctly more pronounced than that observed for 18-crown-6 (triangles in the same Figure 3.2), a behaviour which should reflect the higher degree of steric constraints granted by a spheroidal cavity, offering eight points of interaction, compared to a circular cavity, offering six interaction points. Apparently, cryptand **8**, relaxed to its most stable conformation, provides a tailor-made cavity for K^+.

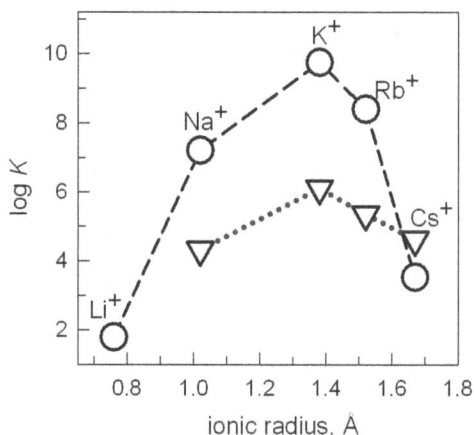

Figure 3.2. Size selectivity in the complexation of alkali metal ions by cryptand **8**. Circles: log K values for the equilibrium $M^+ + L \leftrightarrows [ML]^+$ in MeOH/water 95:5 v/v, at 25°C, M = alkali metal; L = cryptand (**8**).[23] Triangles: log K values for analogous equilibria with L = 18-crown-6 (**5**) in 100% MeOH, at 25°C.

Very interestingly, and this is the most significant contribution of these studies, the cryptand's cavity size and complexation selectivity can be finely adjusted by design. In fact, Lehn and coworkers synthesised a family of analogues of **8**, whose cavity size can be tuned by varying the number of $-(CH_2CH_2O)-$subunits constituting the chains linking the two pivot nitrogen atoms.[24–26] Figure 3.3 illustrates current nomenclature of these bimacrocyclic derivatives, according to which the prototype of cryptands **8** is called 2.2.2-crypt.

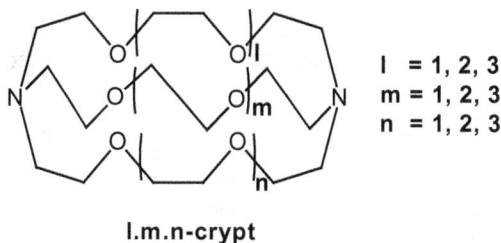

l.m.n-crypt

Figure 3.3. Current nomenclature of cryptands.

In Figure 3.4, log *K* for complexation equilibria in MeOH/water 95:5 v/v involving alkali metal ions and the three cryptands 2.1.1-crypt, 2.2.1-crypt and 2.2.2-crypt have been plotted against metal ionic radius.

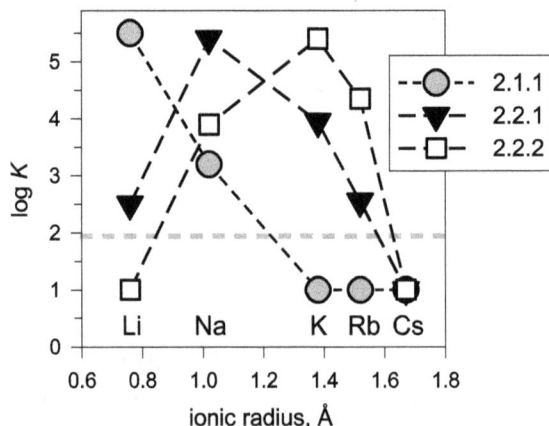

Figure 3.4. Size selectivity in the complexation of alkali metal ions (Li^+ to Cs^+) by 2.1.1-crypt, 2.2.1-crypt and 2.2.2-crypt. Log *K* values for the equilibrium $M^+ + L \leftrightarrows [ML]^+$ in MeOH/water 95:5 v/v, at 25°C, M = alkali metal; L = cryptand: circles, 2.1.1-crypt; triangles: 2.2.1-crypt; squares, 2.2.2-crypt. Log *K* values below the horizontal dashed line could not be determined (log *K* < 2).[23]

Each cryptand displays a sharp size selectivity for a given cation: 2.1.1-crypt for Li^+, 2.2.1-crypt for Na^+ (peak selectivity), 2.2.2-crypt for K^+ (peak selectivity). In other words, each alkali metal ion finds its preferred bimacrocyclic receptor providing a tailor-made cavity. It is thus confirmed that recognition is based on the matching between the radius of the alkali metal ion and the radius of the spheroidal receptor's cavity.

The geometrical nature of selectivity is well illustrated in Figure 3.5, showing the crystal structures of the complexes of Li^+ with 2.1.1-crypt, 2.2.1-crypt, and 2.2.2-crypt. In the most stable complex, [Li(2.1.1-crypt)]⁺, the metal is coordinated by all the available donor atoms (3 O and 2 N), apparently without any serious constraint. In the complex with the immediately higher homologue, 2.2.1-crypt, the metal is still five-coordinate, but one oxygen atom of the cryptand remains uncoordinated,

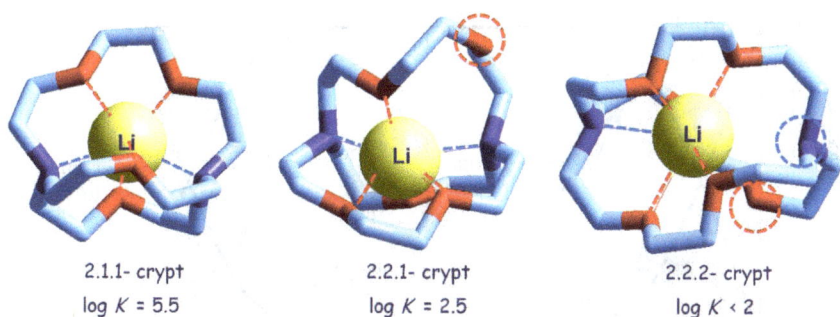

2.1.1- crypt	2.2.1- crypt	2.2.2- crypt
log K = 5.5	log K = 2.5	log K < 2

Figure 3.5. Crystal structures of the Li$^+$ complexes with cryptands 2.1.1-crypt,[27] 2.2.1-crypt,[28] 2.2.2-crypt.[29] Hydrogen atoms and counterions omitted for clarity. Dashed circles indicate oxygen and/or nitrogen atoms not coordinated to the metal.

which demonstrates the difficulty of the ligand to contract its cavity; as a consequence, the binding constant is three orders of magnitude lower than observed for the smaller bicyclic cryptand. Finally, in the 2.2.2-crypt complex, Li$^+$ is lost in the cavity and cannot interact with an oxygen atom and with one tertiary amine group, a circumstance which accounts for its lowest stability.

Indeed, selectivity, in its practical aspects, refers to the possibility of separating two or more analytes, a goal which can be accomplished by using an appropriate ligand. Sodium and potassium go together in many fluids, from seawater ([Na$^+$] = 0.48 M and [K$^+$] = 0.01 M) to blood plasma ([Na$^+$] ~ 0.14 M and [K$^+$] ~ 0.004). Separation of sodium and potassium in water could be, in principle, carried out by using cryptands. In this regard, Figure 3.6(a) simulates an experiment in which an aqueous solution 0.01 M in both NaCl and KCl is titrated with 2.2.1-crypt.

It is observed that, on addition of 2.2.1-crypt, Na$^+$ is complexed first: in particular, on addition of 1 equivalent of ligand, ~85% of [Na(2.2.1-crypt)]$^+$ is formed, whereas the competing complex [K(2.2.1-crypt)]$^+$ is present at ~15%. Material separation could be achieved through a liquid–liquid extraction experiment, by equilibrating the aqueous phase containing the sodium and potassium salts with a chloroform layer containing a lipophilic derivative of 2.2.1-crypt, e.g. the cryptand with a linear –C$_{14}$H$_{29}$ side-chain appended to a carbon atom. Extraction should be facilitated by using poorly hydrated counteranions like picrate. In any case, the

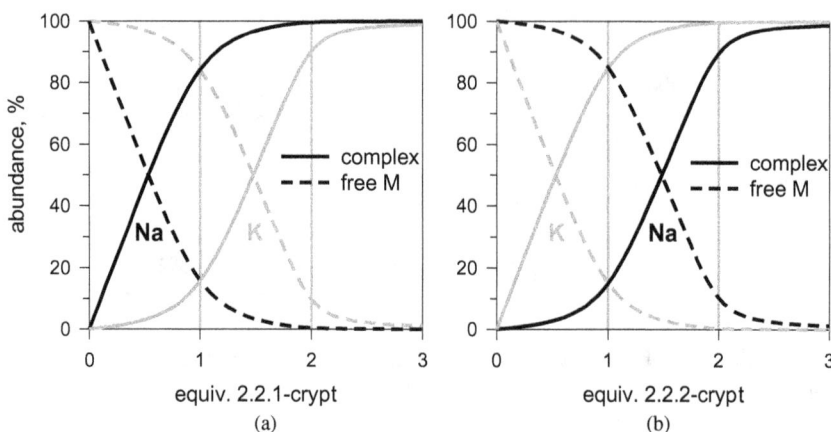

Figure 3.6. Selectivity diagrams. Species present at the equilibrium over the course of the titration of an aqueous solution of 0.01 M in both NaCl and KCl with (a) 2.2.1-crypt, and (b) 2.2.2-crypt. The aqueous solution is buffered at pH > 7 to prevent the protonation of the two tertiary amine groups.

experiment would not be completely satisfactory because exhaustive extraction of [Na(2.2.1-crypt)](picrate) would also involve partial extraction of [K(2.2.1-crypt)](picrate).

Figure 3.6(b) illustrates an experiment in which the same aqueous solution 0.01 M in both NaCl and KCl is titrated with 2.2.2-crypt. The behaviour is reversed with respect to that observed with 2.2.1-crypt: on addition of 1 equivalent of 2.2.2-crypt, ~85% of potassium and ~15% of sodium are complexed, thus not granting exclusive extraction and separation.

From the point of view of the separation of *s*-block cations, things do not go better with alkaline-earth metals with respect to alkali metals. In Figure 3.7(a), log K for complexation equilibria in MeOH/water 95:5 v/v involving Mg^{2+}–Ba^{2+} ions and the three cryptands 2.1.1-crypt, 2.2.1-crypt and 2.2.2-crypt have been plotted against ionic radius. A peak selectivity behaviour is still observed, but to a less pronounced extent than observed for alkali metal ions. This may be due to the fact that the electrostatic contribution to the binding for dipositive alkaline-earth cations is more important that for monopositive alkali metals, a circumstance which attenuates geometrical effects. In any case, 2.2.2-crypt seems especially sensitive to geometrical effects: complex stability increases along the

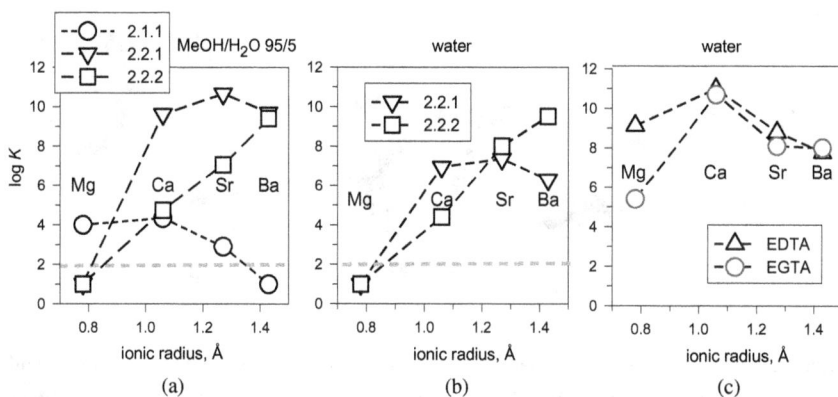

Figure 3.7. Size selectivity in the complexation of alkaline-earth metal ions (Mg^{2+} to Ba^{2+}) by 2.1.1-crypt, 2.2.1-crypt and 2.2.2-crypt. Log K values for the equilibrium M^{2+} + L $\leftrightarrows$ $[ML]^{2+}$: (a) MeOH/water 95:5 v/v[23]; (b) pure water (M = alkaline-earth metal; L = cryptand (log K values below the horizontal dashed line could not be determined, log $K <$ 2)[23]; (c) log K values for complexation equilibria involving EDTA and EGTA (= Y^{4-}); pure water, equilibrium: $M^{2+} + Y^{4-} \leftrightarrows [MY]^{2-}$.[30]

group: $Mg^{2+} < Ca^{2+} < Sr^{2+} < Ba^{2+}$, a sequence which contrasts the decrease of the electrostatic attraction exerted by the metal ions. Moving from MeOH to water does not alter the binding tendencies of 2.2.1-crypt and 2.2.2-crypt, as shown in Figure 3.7(b).

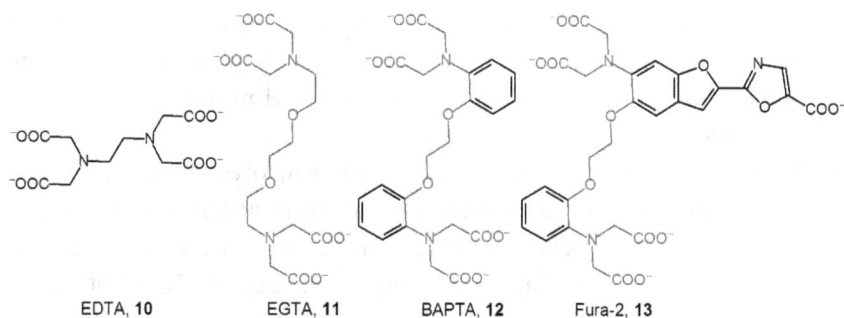

EDTA, **10**   EGTA, **11**   BAPTA, **12**   Fura-2, **13**

Figure 3.7(c) shows log K values for the traditional and widely used ligands for alkaline-earth metals: EDTA (**10**) and EGTA (**11**).[30] EDTA acts as a sexidentate ligand with both Mg^{2+} and Ca^{2+}, a circumstance which has been demonstrated by X-ray diffraction studies on crystalline complexes.

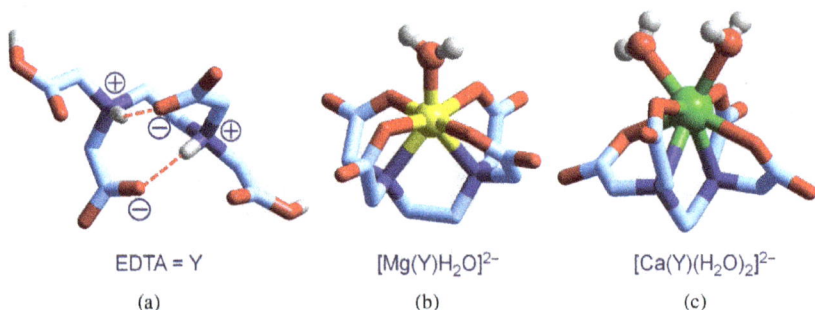

EDTA = Y $[Mg(Y)H_2O]^{2-}$ $[Ca(Y)(H_2O)_2]^{2-}$

(a)   (b)   (c)

Figure 3.8. Crystal and molecular structures of (a) EDTA in its zwitterionic form,[31] and of its magnesium (b)[32] and calcium (c)[33] complexes. Hydrogen atoms, except those of metal-coordinated water molecules, as well as counteranions have been omitted for clarity.

In particular, Figure 3.8(a) shows the structure of the uncomplexed ligand in its zwitterionic form. The conformational arrangement is stabilised by intramolecular hydrogen bonding interactions involving carboxylate oxygen atoms and ammonium groups. In the magnesium complex (Figure 3.8(b)), four carboxylate oxygen atoms, two tertiary amine nitrogen atoms and one water molecule are bound to the metal. The bigger calcium ion coordinates two additional water molecules (Figure 3.8(c)).

EDTA resembles cryptands in that it contains two nitrogen atoms and four oxygen donor atoms (like 2.2.2-crypt), but in addition the oxygen atoms detain a formal negative charge, which enhances the electrostatic contribution to the metal–ligand interaction. Diaminopolycarboxylates, of which EDTA is the prototype, have been involved in the separation and selective determination of s block metal ions. A particular issue refers to the determination of Ca^{2+} in the cell. Calcium is involved in several physiological functions, and the variation of its concentration monitors the development of the function: muscle concentration, nerve communication, hormone secretion, immune activation. Intracellular $[Ca^{2+}]$ at rest is 10^{-7} M and under stimulus reaches 10^{-6}–10^{-5} M. Any analyte at 10^{-7} concentration can be detected by an appropriate fluorescent receptor and, in cell physiology studies, revealed in real time and real space with a fluorescence microscope. The problem is that in the cell Ca^{2+} has to compete for the receptor with Mg^{2+}, which is present in over high concentration,

10^{-3} M. Thus, in order to achieve successful recognition, the envisaged receptor should present a binding constant for Ca^{2+} more than four orders of magnitude higher than for Mg^{2+}.

EDTA, in its tetradeprotonated version, Y^{4-}, forms with Ca^{2+} a more stable complex than with Mg^{2+} (log K 11.0 and 9.1, respectively), but the difference in stability is too small to guarantee selective recognition inside the cell. A more favourable situation is provided by EGTA (**11**), which contains two additional ethereal oxygen atoms in its donor set. It is not clear whether these two oxygen atoms are involved in coordination. In any case, log K for Ca^{2+} is 11.0 and for Mg^{2+} is 5.2, a difference which affords selective binding of 10^{-7} M calcium even in the presence of 10^{-3} M magnesium.

Roger Y. Tsien, in the 1980s, was looking for a fluorimetric molecular sensor for determining intracellular Ca^{2+} and took inspiration from EGTA.[34] In particular, he synthesised system **12** (BAPTA, acronym of **B**is(o-**A**mino**P**henoxy)ethane-*N,N,N′,N′*-**T**etra**A**cetic acid), in which each of the two ethylene chains linking a nitrogen atom and an oxygen atom has been replaced by an ortho-xylyl fragment. As the aniline group is distinctly less basic than its amine counterpart, BAPTA nitrogen atoms are much less affected by pH changes compared to those of EGTA. In particular, at physiological pH, BAPTA's nitrogens are not protonated, whereas those of EGTA are. The less basic nature of the aniline nitrogen atom reduces the stability of the Ca^{2+}-BAPTA complex compared to EGTA (log $K = 7.0$ and 11.0, respectively), but this also happens for the Mg^{2+} complex (log $K = 1.8$ and 5.2, respectively). Thus, selectivity is maintained (Δlog $K > 5$). BAPTA displays an emission band centred at 363 nm, with no fine structure and with quantum yield of 2.7%. On binding of Ca^{2+}, the emission intensity is reduced to 1/3 without alteration of band shape and peak wavelength.

A few years later, Tsien introduced an improved class of fluorescent indicators of calcium, whose prototype is Fura-2 (**13**).[35] Fura-2 maintains the same structural formula of BAPTA, but possesses a powerful chromophoric fragment (2-furan-2yl)oxazole-carboxylate) integrated to one of the phenyl fragments. Fura-2 shows an absorbance maximum in the UV region (362 nm, molar extinction coefficient $\varepsilon = 27000$) and an emission maximum in the visible region (518 nm) with a quantum efficiency of

23%. On formation of the 1:1 complex with calcium, the absorption band of Fura-2 is blue-shifted (335 nm) with an increase in its intensity ($\varepsilon =$ 33000 M^{-1} cm^{-1}), while the emission band undergoes a moderate blue shift (to 510 nm) and a dramatic increase of the quantum yield (49%).

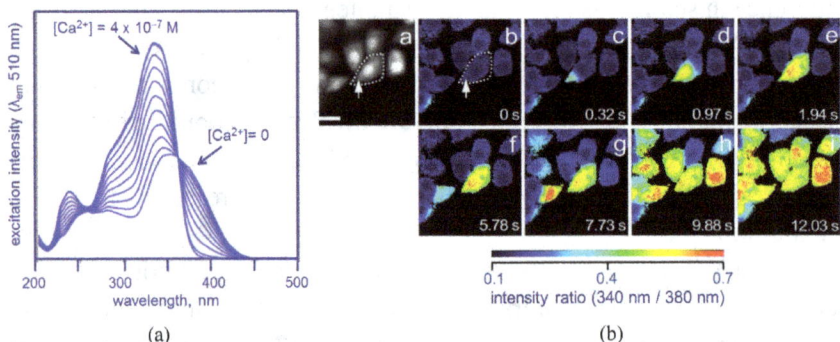

Figure 3.9. (a) Excitation spectra taken over the course of the titration of an aqueous solution of Fura-2 (**13**) with Ca^{2+}; (b) mechanical-stimulation-induced Ca^{2+} responses in a single HSY-EA1 cell over a period of 12 s. The Fura-2-loaded cell (indicated by a dashed line in frames **a** and **b**) has been mechanically stimulated with a glass micropipette. Frame **a**: Fura-2 fluorescence image at 380 nm excitation. Frames **b**-to-**i**: images obtained from excitation intensity ratio (340/380 nm). Arrow, position of mechanical stimulation. Pictures taken from Ref. [36].

Figure 3.9(a) shows the excitation spectra obtained on addition of Ca^{2+} to a solution of Fura-2. Noticeably, Fura-2 displays unique spectral features, very useful for application: (i) the high quantum efficiency, which allows determination of calcium in real time and real space using a fluorescence microscope (e.g. localisation of Ca^{2+} inside a single muscle cell following stimulation); (ii) the well-detectable separation of the excitation/emission bands of Fura-2 and of its calcium complex, which makes Fura-2 the ideal *ratiometric* indicator: in particular, calcium concentration can be determined from the *ratio* of the fluorescence intensities determined at two distinct wavelengths, that of the excitation band of the complex (340 nm) and that of the excitation band of the metal-free indicator (380 nm). Figure 3.9(b) shows a practical example of the variation of the Ca^{2+} concentration inside a single HSY-EA1 cell following mechanical stimulation by a glass micropipette of a Fura-2-loaded cell.[36] The dashed lines in frames **a** and **b** isolate the cell under investigation, while the white arrows indicate the position of the stimulation. The Ca^{2+} response begins at the stimulated region and spreads throughout the cell within 2 seconds (frames **b-to-e**)). Subsequent Ca^{2+} responses propagate from the mechanically stimulated cell to neighbouring cells as an intercellular Ca^{2+} wave (frames **f-to-i**).

The development of fluorescent indicators for calcium has not involved more sophisticated ligands as cryptands, a prospect which cannot be excluded for the future. Synthetic complexity should not discourage the design of calcium-responsive fluorescent indicators based on cryptands, in view of the astronomical prices of Fura-2 and its analogues: 1 mg of Fura-2 pentasodium salt costs ~€200.

4

Competitive Hosts for Alkali Metal Ions: Spherands and Derivatives

Pedersen introduced crown ethers in 1967, thus unfolding the rich field of coordination chemistry of s block metal ions. The family of ligands containing ethereal oxygen atoms suitable for coordination of alkali and alkaline-earth cations was expanded two years later by Lehn with the synthesis of cryptands. Pedersen and Lehn were awarded with the Nobel Prize in Chemistry in 1987. The third recipient of the most prestigious award in chemistry in the same year was Donald Cram (1919–2001), Professor at the University of California at Los Angeles. Cram no wonder did extensive studies on the design of multidentate ligands for alkali metal ions. In particular, in 1985, he reported the synthesis of the cyclic molecule **14**, containing a system of six linked methoxybenzene (anisyl) fragments.[37]

Molecule **14** has been clearly inspired by crown ethers and, like 18-crown-6, it contains six ethereal oxygen atoms. Compared to Pedersen's crowns, which contain ethylene spacers, molecule **14**, whose carbon skeleton is made of sp^2 atoms, is expected to be much more rigid.

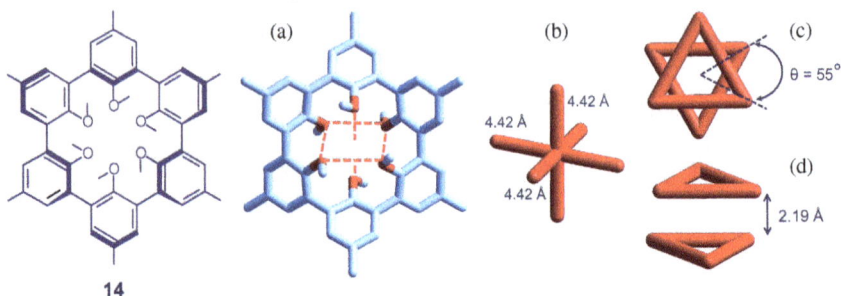

14

Figure 4.1. Spherand **14**: (a) the crystal structure of *spherand* **14**.[37] Dashed red lines define the octahedron, whose vertices are occupied by the six anisyl oxygen atoms; (b) lines obtained by joining the opposite vertices of the slightly distorted octahedron constituted by the oxygen atoms of the spherand (length: 4.42 Å); (c) three oxygens on one side of the octahedron have been linked to give an equilateral triangle and the same for the opposite three oxygens (top view); the angle formed by the two staggered triangles, twist or torsion angle, is 55°; in a regular octahedron, $\theta = 60°$; (d) lateral view.

Looking at the apparently planar structural formula the name given by Cram to **14**, *spherand*, may be surprising. However, CPK molecular models and, more convincingly, the crystal structure showed that the methoxyl groups, in order to minimise mutual steric repulsions, are not coplanar, but show an alternating 'up and down' arrangement.[37] In particular, the plane containing each phenyl ring is rotated by 52° with respect to the plane of each neighbouring ring. A relevant consequence is that the six donor oxygen atoms are placed at the corners of a nearly regular octahedron, as illustrated in Figure 4.1(a). The current name of *spherand* could be explained by considering the fact that an octahedron can be inscribed in a sphere.

The octahedral donor set of oxygen atoms is suitable for the coordination of *s* block metal ions of appropriate size. Figure 4.1(b) suggests that a metal ion, in order to be perfectly included in the O6 cavity of the spherand, should be placed at 2.21 Å from each oxygen atom (= 4.42/2 Å). As mentioned in Chapter 1, the 'natural' distance M–O of alkali cations can be tentatively estimated by looking at complexes with the terdentate ligand *bisdiglyme*, in particular, considering the bond involving the central oxygen atom of the polyether. In this way, the following M–O distances are obtained: Li$^+$, 2.11 ± 0.01 Å;[38] Na$^+$, 2.41 ± 0.02 Å; K$^+$, 2.87 ± 0.06 Å.

It is observed that the Li–O distance of the [Li(bisdiglyme)$_2$]$^+$ complex (2.11 Å) is only slightly lower than that required by the empty spherand (2.21 Å), a situation which should ensure a very easy inclusion in the cavity. On the other hand, the Na–O distance (2.41 Å)[13] observed in the [Na (bisdiglyme)$_2$]$^+$ complex would indicate that metal inclusion in the spherand should involve an energy cost in terms of conformational rearrangement. Moreover, metal inclusion can be reasonably excluded for potassium, whose 'natural' K–O distance (2.9 Å)[14] is substantially larger than that tolerated by the spherand.

Predictions based on the structural data were fulfilled by extraction experiments: a CHCl$_3$ solution 2×10^{-3} M of the spherand was equilibrated with an equal volume of an aqueous solution of an MX salt (M = Li, Na, K, Cs; X = Cl, Br). Only LiX and NaX salts were extracted, and when in the organic layer, metal cations were fully complexed by the spherand. Aqueous KX and CsX were not extracted, even if present at 6 M concentration.[39]

Quantitative information on the stability of metal complexes of **14** (*spherates*, one should say along the line of cryptates' nomenclature, but Cram had preferred *spheraplexes*) came from equilibrium studies. Pertinent investigations were carried out in a water-saturated CHCl$_3$ solution at 25°C.[40] The solubility of water in chloroform at 25°C is 0.059 g/100 g CHCl$_3$, which corresponds to 2.1×10^{-2} M, a concentration non-negligible because it is of the same order of magnitude as that of the investigated analytes. Chloroform, a poorly polar solvent (1.15 D), was chosen in order to guarantee solubility of the highly lipophilic spherand **14**. To ensure full solubility in chloroform, alkali metal ions were dissolved as picrates. Complexation equilibria were investigated through ^{1}H NMR titrations and the log K values for the formation of the 1:1 complexes were determined: they are >23 for Li$^+$ and 14.0 for Na$^+$. It has to be observed that spherand **14** forms with lithium the most stable complex ever observed in the coordination chemistry of s block metal ions and that it shows an unprecedented selectivity for lithium over sodium (and over heavier alkali cations, which are not complexed at all).

Crystal structures of the two spherates, shown in Figure 4.2,[37] could help elucidate these points.

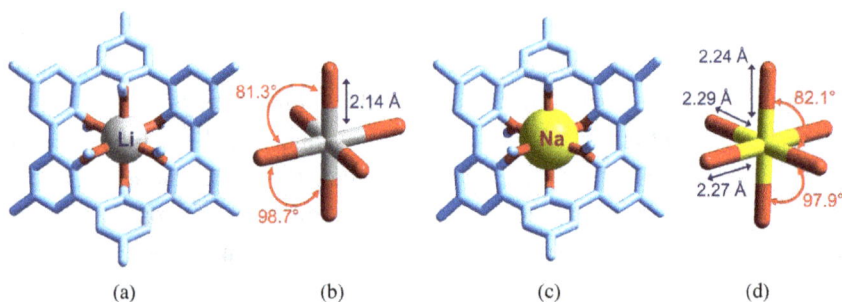

Figure 4.2. Structure of the spherate complexes: (a) the crystal structure of the [Li(L)] Cl complex salt (L = **14**)[37]; (b) octahedral arrangement of the six ethereal oxygen atoms bound to Li[+]; (c) the crystal structure of the [Na(L)]CH₃OSO₃·CH₂Cl₂ complex salt[37]; (d) octahedral arrangement of the six ethereal oxygen atoms bound to Na. Hydrogen atoms, counterions and solvating molecules have been omitted for clarity.

In particular, the Li[+] cation appears well included in the spherand's cavity (Figure 4.2(a)), coordinated by the six oxygen atoms according to a nearly perfect octahedral geometry: the Li–O distance is the same for all bonds (2.11 Å), is coincident with the 'natural' value and is only slightly lower than that expected for the empty ligand. In particular, the structure of the spherand framework in the complexes is substantially the same as that of the metal-free ligand. The situation of the sodium complex appears less favourable: Na–O distances range from 2.24 to 2.29 Å, i.e. values appreciably smaller than the 'natural' distance of Na–O bond (~2.4 Å). Apparently, the relatively small spherand's cavity has to spend a substantial energy cost to accommodate the relatively large Na[+] cation, which, in turn, cannot profit in full from the coordination of the six oxygen atoms. Such an energy cost roughly amounts to more than 12 kcal mol[−1], i.e. the difference of $\Delta G°$ values of complexation (<31 kcal mol[−1] for lithium and −19.1 kcal mol[−1] for sodium). There is no way to include potassium into the spherand's cavity: the 'natural' K–O distance (~2.9 Å) largely exceeds that tolerated by the spherand.

The extreme stability of the lithium spherate complex illustrates well Cram's principle of *preorganisation*, which states: "the more highly hosts and guests are organized for binding and for low solvation prior to their complexation, the more stable will be their complexes". In the language of biological chemistry, the 'host' is the 'receptor' and the 'guest' is the 'substrate'; in the more restricted language of coordination chemistry, the 'host' is the 'ligand' and the 'guest' is the 'metal centre'. Indeed, spherand **14** alone exhibits a structure already arranged for the coordination of a metal ion of appropriate size with the lone pairs of its oxygen donor atoms properly oriented towards the centre of an octahedral cavity. Thus, metal coordination eliminates the mutual electrostatic repulsions between the convergent lone pairs. As a further advantage, the rigidity of the molecular skeleton prevents the host from assuming a variety of conformations, which minimises the loss of conformational energy during complexation. Finally, the compact and lipophilic nature of the cavity excludes or reduces at minimum the interaction of solvent molecules with the convergent oxygen lone pairs. In conclusion, the energy term associated with the organisation of the ligand to guarantee the most favourable conformation for metal coordination is spent during the ligand's synthesis and not during complexation. The title of the opening paper of the series was just "Spherands, the First Ligand Systems Fully Organized during Synthesis Rather than during Complexation".[37]

The term 'preorganisation' had a notable success in the chemical lexicon and is currently used by authors to describe recognition studies. It is reminiscent of the term 'reorganisation', introduced by Marcus in the theory of electron transfer (eT)[41]: the less important the reorganisation of the molecular framework and of the solvating molecules when moving from the precursor to the successor, the faster the eT process. Cram's preorganisation refers to thermodynamics ($\Delta G°$), Marcus' reorganisation refers to kinetics ($\Delta G^{\ddagger}$).

An obvious development of spherand **14** was macrocycle **15**, made of eight covalently linked anisyl subunits, thus providing a cavity large enough to also include heavier alkali metal ions.[42] Figure 4.3(a) shows the crystal structure of **15**. It is observed that the eight benzene rings, like in spherand **14**, are arranged according to an 'up and down' sequence with dihedral angles ranging from 60° to 90°. It shows that the eight anisyl oxygen atoms are not coplanar, but are placed at the corners of a definite solid: a square antiprism. In particular, Figure 4.3(b) shows two opposite tetragonal faces of the polyhedron: one quadrilateral has been obtained by linking the oxygens up, the other by linking the oxygens

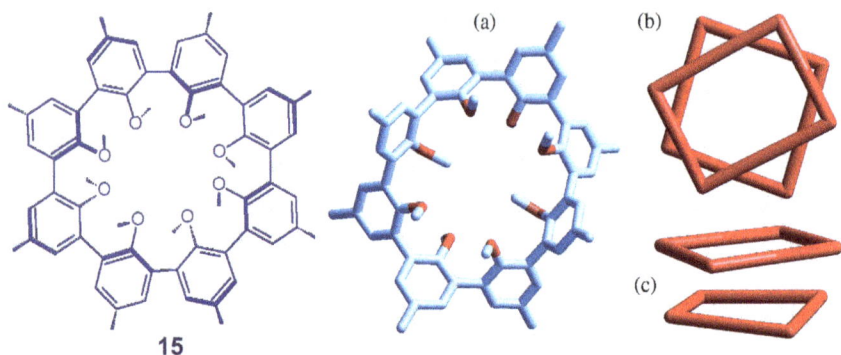

Figure 4.3. Structure of cavitand **15**: (a) the crystal structure of L·CH$_2$Cl$_2$ (L = **15**)[43]; hydrogen atoms and solvating molecule have been omitted for clarity; the macrocycle shows an alternating 'up and down' sequence of covalently linked anisyl units; (b) the quadrilateral obtained by linking the oxygen atoms *up* and the quadrilateral obtained by linking the oxygen atoms *down*: the 4 + 4 oxygens are placed at the corners of a distorted square antiprism; the two quadrilaterals represent two opposite faces of the solid; (c) lateral view of the two quadrilaterals.

down. The square antiprism provides the most convenient geometrical arrangement for eight-coordination. For macrocycle **15**, Cram proposed the name *cavitand*. In a general sense, a cavitand is a synthetic organic compound that contains an enforced cavity large enough to accommodate simple molecules or ions.[44] Thus, spherand **14** must be considered as a distinguished member of this broad class of hosts.

Noticeably, cavitand **15** forms complexes with all alkali metal ions. Figure 4.4 shows a plot of log K values of its 1:1 complexes, determined in water-saturated chloroform, at 25°C, vs ionic radius.[42] For comparison, log K values for the complexes of Li$^+$ and Na$^+$ with spherand **14** are also shown in the figure. It must be preliminarily observed that lithium and sodium complexes of spherand **14** are exceedingly more stable than complexes of cavitand **15** with all alkali metals. Moreover, cavitand complexes (named *caviplexes* by the authors) show a size-dependent selectivity behaviour: stability increases with the increasing metal ionic radius. Cs$^+$ forms the most stable caviplex. In particular, [Cs(L)]$^+$ is four orders of magnitude more stable than [Li(L)]$^+$ for L = **15**, but this selectivity factor is nothing when compared to that exhibited by [Li(L)]$^+$ over [Na(L)]$^+$ for L = **14**.

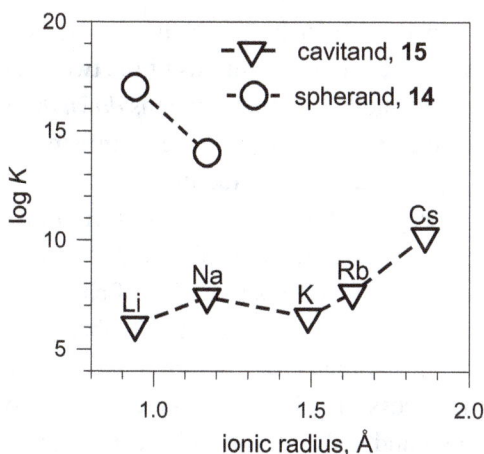

Figure 4.4. The relative stability of alkali metal complexes of cavitand **15** and spherand **14**. Log *K* values of 1:1 complexes determined in water-saturated chloroform at 25°C.[42]

Figure 4.5(a) shows the crystal structure of the $[Cs(L)]ClO_4 \cdot CH_2Cl_2$ complex salt (L = **15**).[43]

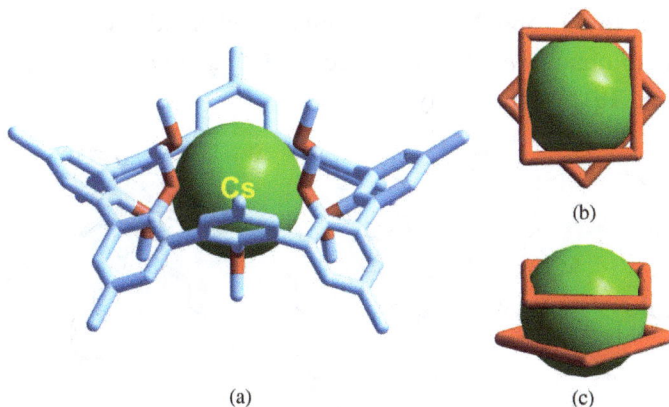

Figure 4.5. The caesium caviplex: (a) the crystal structure of the $[Cs(L)]ClO_4 \cdot CH_2Cl_2$ complex salt (L = **15**)[43]; ClO_4^- and CH_2Cl_2 have been omitted for clarity; (b) top view of the quadrilaterals whose corners are the oxygens up and the oxygens down of cavitand **15**: the reciprocal orientations of the two polygons, staggered by ~45° indicate a square antiprism coordination polyhedron; (c) lateral view.

The caesium ion is well included in the cavity, bound to the eight anisyl oxygen atoms. The relative positions of the two quadrilaterals made by the four oxygens *up* and by the four oxygens *down* define the coordination polyhedron well: a nearly regular square antiprism (see Figures 4.5(b) and 4.5(c)). Such a polyhedron is not too different from that present in the metal-free host (see Figures 4.3(b) and 4.3(c)), a circumstance which fulfils the requirements of the principle of preorganisation. Therefore, the relatively low stability of caviplexes (e.g. that of caesium) with respect to spheraplexes (e.g. that of lithium) should be ascribed to something different from a mere conformational effect. Cram suggested that, in the smaller macrocycle **14**, the access of solvent molecules to the cavity is sterically denied. On the other hand, solvent molecules can occupy the cavity of the larger macrocycle **15**. It has been mentioned that in a water-saturated chloroform solution, H_2O molecules are present at 2×10^{-2} M concentration level, which corresponds to several water molecules per molecule of host, enough for filling the cavity. Thus, the complexation by cavitand should involve the dehydration of the host, an endothermic process which may substantially reduce the free energy of complexation and the binding constant. In other words, cavitand **15** is not preorganised to prevent solvation, whereas spherand **14** is.

In order to extend and modulate the binding tendencies of spherand **14** (limited to Li^+ and Na^+), Cram mixed spherands with crowns and cryptands to obtain *hemispherands* (e.g. **16**)[44] and *cryptahemispherands* (e.g. **17** and **18**),[45] respectively.

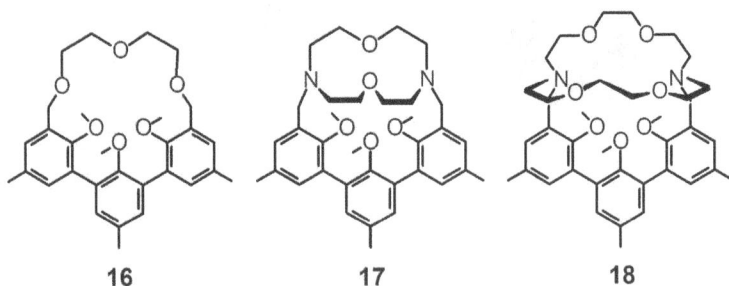

16   17   18

Hemispherands and cryptahemispherands are less rigid and less sterically constrained than spherand **14**, but maintain a pronounced degree of preorganisation, which affords the formation of stable complexes.

In Figure 4.6, log K values for the complexation equilibria involving hemispherand **16** and cryptahemispherands **17** and **18** are compared to the values determined for chosen crowns and cryptands under the same conditions, i.e. water-saturated chloroform at 25°C.[44,45]

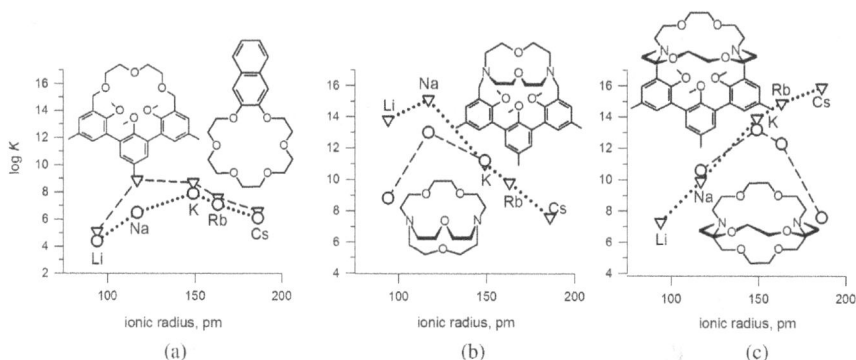

Figure 4.6. Log K values of complexation equilibria (1:1 complexes) in water saturated chloroform at 25°C; (a) log K values for hemispherand (**16**, triangles) are compared with those of naphtho-18-crown-6, circles; (b) log K values for 1.1-cryptahemispherand (**17**, triangles) are compared with those of 2.1.1-crypt, circles; (c) log K values for 2.2-cryptahemispherand (**18**, triangles) are compared with those of 2.2.2-crypt, circles.[44,45]

In Figure 4.6(a), the coordinating properties of hemispherand **16**, which possesses a macrocyclic structure and contains six ethereal oxygen atoms, are compared with those of the classical Pedersen's hexadentate macrocycle 18-crown-6, to which a naphtho substituent has been attached in order to impart solubility in chloroform. The hemispherand forms with all alkali metals complexes slightly more stable than those of the crown with a similar selectivity pattern. The 1.1-cryptaspherand (**17**) is compared to the Lehn's 2.1.1-crypt in Figure 4.6(b). Both ligands show selectivity in favour of sodium. Finally, the 2.2-cryptahemispherand (**18**) forms with Li$^+$, Na$^+$ and K$^+$ complexes of the same stability as 2.2.2-crypt (Figure 4.6(c)). However, while the cryptand displays selectivity towards K$^+$, the stability of the cryptahemispherand complexes keeps increasing up to Cs$^+$. In conclusion, the insertion of aliphatic moieties in the framework of the spherand makes rigidity and preorganisation vanish and the corresponding complexes show stability and selectivity comparable to those of crown ethers and cryptands.

Crowns and cryptands find application in many areas of chemistry, biochemistry and material science, an opportunity not shared by spherands and their derivatives. This may be due to their tedious multistep synthesis and perhaps to the fact the crowns and cryptands came first. A unique and unbeatable property of spherand **14** is its capability to sequester specifically Li^+ in the presence of any other metal ions. Such a feature could be of interest in the recovery of lithium from spent batteries, which is at present carried out by using less efficient (but also cheaper) commercial ligands.

Host, guest and the unavoidable goodness of human beings

A *host* is one that receives or entertains a guest socially, commercially, or officially. Conversely, a *guest* is one that is received or entertained by a host. Both words are derived from Latin *hostis*, enemy, a common etymology which may be surprising at first sight. Italians are more complicated: they say 'ospite' for both 'host' and 'guest', which often creates misunderstandings in everyday language. The same happens in the French tongue with *hôte*. 'Ospite' is derived from Latin *hostis potis*, the most powerful *enemy*. This means, or meant 2700 years ago, when Latin civilisation developed, that, because you are the most powerful of my *enemies*, I do not want to fight you or kill you, but, on the contrary, I welcome you and I wish that you profit from my friendship and from the goods my house can offer you. And, in turn, you must trust me because I am your powerful enemy. Why? Because of the two options — (1) host and guest fight and one kills the other; (2) host and guest become friends and enjoy life at the host's home — the second one is more pleasant and also more proficient in an evolutionary sense. Thus, the Latin sentence *Homo homini lupus* (a man is a wolf to another man), which had a broad acceptance in Western culture since Plautus to Hobbes and Freud, must be considered inconsistent and unnatural. Human beings are unselfish and inclined to cooperation. If a man were a wolf to other men, the human race would not have developed and would have disappeared thousands and thousands of years ago.

Hostis and *hostis potis* have generated a variety of words in current use: hospital, hotel, hostess. Also 'hostile', which, nevertheless, refers to the inhuman acceptation of *hostis*.

5

The Coordination Chemistry of Ammonium and Oxonium

NH_4^+ is the most prominent of inorganic polyatomic cations, constituted by non-metal atoms and showing a regular tetrahedral structure. It is conventionally considered an additional alkali metal ion essentially because all its salts are soluble, a unique feature of alkali metal ions. In terms of size, the ammonium cation has an ionic radius ($r = 148$ pm) very close to that of K^+ ($r = 149$ pm). If crown ethers and cryptands opened up the field of coordination chemistry of alkali metal ions, what about the interaction of these multidentate ligands with ammonium? Indeed, ammonium shares a tendency with alkali cations to form stable complexes with crown ethers, cryptands and other cyclic and polycyclic polyethers.

The 1:1 complex of ammonium with 18-crown-6 has been isolated in a series of salts. Figure 5.1 shows the crystal structure of the complex salt $[(NH_4)(18\text{-crown-}6)]NCS \cdot H_2O.$[46] It can be observed from Figure 5.1(a) that the ammonium cation lies distinctly above the mean plane of the six ethereal oxygen atoms of the crown. Such a structural organisation is different from that observed in the corresponding 1:1 complex of K^+, in

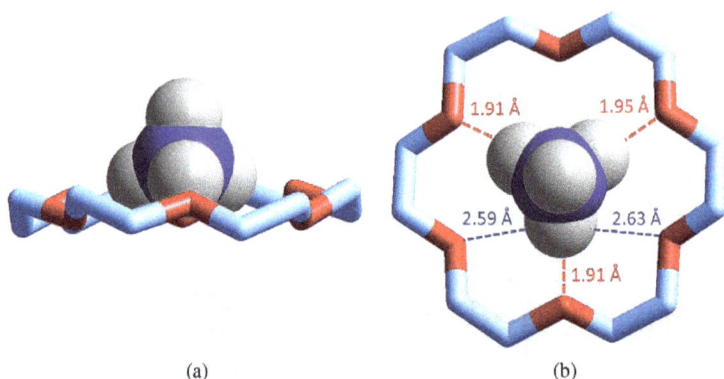

Figure 5.1. Crystal structure of the [(NH$_4$)(18-crown-6)]NCS·H$_2$O complex salt.[46] Thiocyanate counterion, solvating molecule and C–H hydrogen atoms have been omitted for clarity; (a) lateral view; (b) top view. Only three hydrogen atoms of ammonium establish H-bond interactions with the crown: each one of the three hydrogens forms a major H-bond with a facing ethereal oxygen (red dashed line) and establishes two minor interactions with the two oxygen atoms adjacent to the facing one (blue dashed line, illustrated for one hydrogen only).

which the metal is perfectly coplanar with the O6 mean plane (Figures 1.4 and 1.5) and establishes especially strong interactions with the ligand, electrostatic in nature. The bonding of the ammonium cation to the macrocycle results from a different type of interaction: hydrogen bonding. In particular, only three hydrogen atoms of NH$_4^+$ interact with the crown ether: each one of these hydrogen atoms establishes a major interaction with a facing ethereal oxygen atom and two minor interactions with the two oxygen atoms nearby to the facing one, as illustrated in Figure 5.1(b).

Equilibrium studies in water have shown that the [K(18-crown-6)]$^+$ complex is one order of magnitude more stable than the [(NH$_4$)(18-crown-6)]$^+$ complex (log K values = 2.1 and 1.1, respectively).[47] The different stability is a consequence of the diverse nature of the interaction: potassium has the correct size to give a coplanar arrangement, thus establishing the most energetic coulombic interactions. On the other hand, of the four polarised N–H fragments of tetrahedral NH$_4^+$, only three are involved in bonding interactions, whereas one remains unutilised. Failure of ammonium to profit from all its bonding tendencies may account for the lower stability of its complex with 18-crown-6 compared to the potassium analogue (one order of magnitude).

The tetrahedral NH_4^+ cation should benefit to a larger extent from the interaction with a receptor providing a tridimensional cavity, like, for instance, 2.2.2-crypt. Figure 5.2 shows the crystal structure of the complex salt $[NH_4(2.2.2\text{-crypt})]Cl\cdot3.5H_2O$.[48]

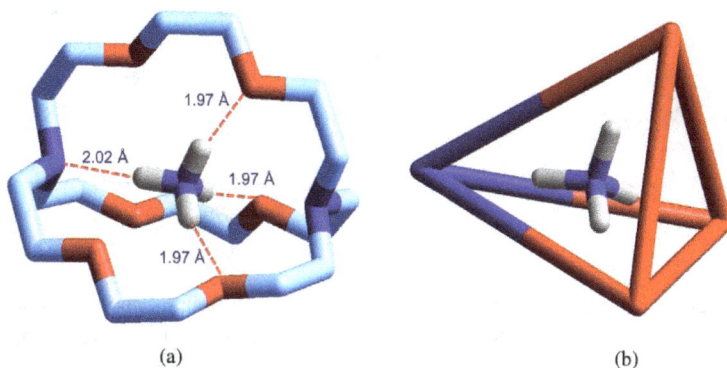

Figure 5.2. Crystal structure of $[NH_4(2.2.2\text{-crypt})]Cl\cdot3.5H_2O$ complex salt.[48] Chloride counterion and C–H hydrogen atoms have been omitted for clarity: (a) dashed red lines emphasise the directional hydrogen bonding interactions involving the four N–H fragments of ammonium, three ethereal oxygens and one of the two bridgehead nitrogens; (b) the tetrahedral ammonium cation is included in the tetrahedron, whose apices are the 3 oxygen atoms and the nitrogen atom that establish with NH_4^+ major H-bond interactions.

The ammonium cation is well included in the cryptand's cavity and, as evidenced by the dashed lines in Figure 5.2(a), it establishes four major hydrogen bonding interactions with four H-bond acceptor atoms of the receptor. In particular, three N–H fragments point towards a facing oxygen atom, one N–H fragment points towards one of the two tertiary nitrogen atoms. Moreover, each hydrogen atom establishes 2–3 minor H-bond interactions with more distant O/N atoms. The stability of the H-bond arrangement is also demonstrated by the fact that no proton of NH_4^+ is transferred to the proximate bridgehead nitrogen atom, which, being a part of a tertiary amine group, should present an intrinsically higher basicity than NH_3.

Cram determined the association constants of ammonium and alkali cations with crowns, cryptands and other polycyclic polyethers in his

preferred solvent: water-saturated chloroform, at 25°C.[49] Significant
results are illustrated in Figure 5.3.

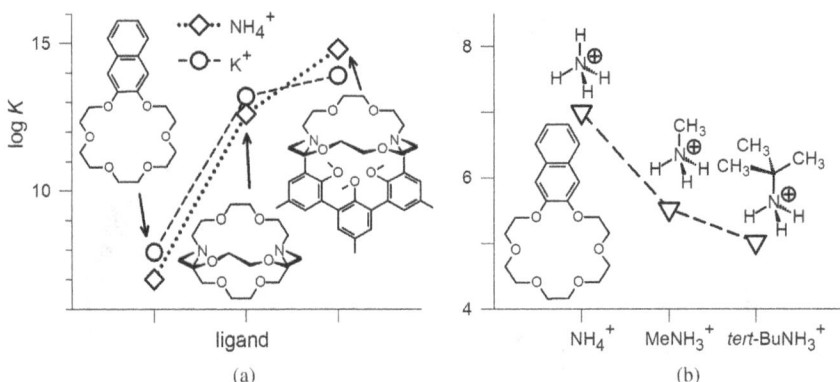

Figure 5.3.   (a) log K values of the association constants of ammonium and potassium
with selected crowns, cryptands and cryptahemispherands in water-saturated chloroform,
at 25°C;[49] (b) log K values of the association constants of ammonium and alkylammonium
ions with 3,4-naphtho-18-crown-6 in water-saturated chloroform, at 25°C.[50]

Figure 5.3(a) shows that the stability of the ammonium complex with
2.2.2-crypt greatly exceeds that of the corresponding complex with
3,4-naphtho-crown, a behaviour which reflects the formation of four
major hydrogen bonds with the tridimensional receptor instead of the
three major H-bonds established with the bidimensional polyether. The
2.2-cryptahemispherand (**18**) forms an even more stable complex with
ammonium. It is possible that the cryptahemispherand offers to NH_4^+ a
better set of H-bond acceptor atoms than the cryptand, a situation which
can be favoured by the rigidity of the anisyl-containing moiety. In any
case, it is observed that ammonium and potassium, even if bound to the
ligand through interactions of a different nature, form complexes of comparable stability.

However, 2.2.2-crypt and 2.2-cryptahemispherand are not tailor-made
ligands for NH_4^+. Ammonium is tetrahedral and is expected to profit from
the inclusion in a tetrahedral receptor. From Figure 5.2(b), it can be seen
that the tetrahedron obtained by linking the atoms involved in major

hydrogen bonding interactions (three ethereal oxygens and one tertiary nitrogen) is noticeably distorted. Hydrogen bond has a directional character and such a deviation from regularity prevents the formation of strong bonds.

A receptor containing a perfect tetrahedral cavity suitable for directional interaction with ammonium has been designed by Lehn,[51] and its structural formula is shown in Figure 5.4.

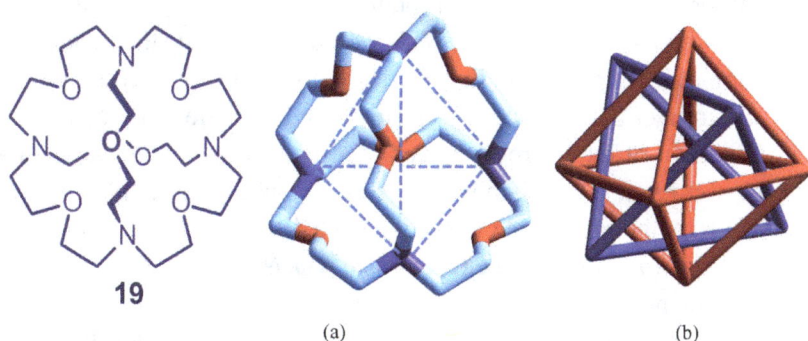

19

(a)   (b)

Figure 5.4. A spherical cryptand: (a) the calculated structure of macrotricycle **19**; tertiary nitrogen atoms are placed at the corners of a regular octahedron, whose edges are indicated by blue dashed lines; ethereal oxygen atoms are positioned at the corners of a regular octahedron (not represented); (b) the nitrogens' tetrahedron inserted in the oxygens' octahedron.

Receptor **19** contains four tertiary amine nitrogen atoms placed at the corners of a tetrahedron (see the calculated structure in Figure 5.4(a)), which are linked by six $-CH_2CH_2OCH_2CH_2-$ chains, laid on the edges of the tetrahedron. The interesting point is that the six ethereal oxygen atoms end up being positioned at the corners of an octahedron. Thus, the spherical cryptand offers to cations a variety of opportunities to satisfy their geometrical requirements.

Ammonium forms with **19** an especially stable 1:1 complex, whose log K value in water at 25°C is 6.1.[51] The crystal structure of the complex salt $[NH_4(L)]I \cdot H_2O$ (L = **19**) has been determined.[52] It is shown that ammonium points each one of its four N–H fragments towards one of the four nitrogen atoms of the receptor, to give a collinear N–H$\cdots$N pattern.

The N---N distances range over the 3.05–3.17 Å interval, values larger than observed for instance in the ammonium complex of 2.2.2-crypt (2.92 and 2.97 Å). This suggests that the cavity offered by the spherical cryptand may be slightly larger than demanded by NH_4^+. Moreover, each hydrogen atom of NH_4^+ establishes with two proximate ethereal oxygens of **19** two minor bent N–H$\cdots$O interactions. The total of four major collinear N–H$\cdots$N and 12 minor bent N–H$\cdots$O hydrogen bonding interactions contribute to the unusually high stability of the complex. Potassium and rubidium form with the spherical cryptand distinctly less stable complexes (log $K = 3.4$ and 4.2, respectively). This would demonstrate that, under the same geometrical conditions, the hydrogen bonding interaction (directional) is stronger than the merely electrostatic interaction (adirectional).

Substitution of the hydrogen atoms of ammonium with alkyl/aryl substituents gives rise to a large family of organic cations. Figure 5.3(b) compares log K values for the complexation equilibria of 3,4-naphtho-crown with NH_4^+, $CH_3NH_3^+$ and *tert*-BuNH$_3^+$,[50] a circumstance which allows to evaluate steric effects on complexation of primary alkylammonium cations. It is observed that the log K value diminishes along the series $NH_4^+ >$ MeNH$_3^+ > $ *tert*-ButNH$_3^+$, a sequence which can be perfunctorily ascribed to repulsive effects between the ammonium substituent and the ligand's framework. Available crystal structures allow to formulate a more subtle explanation of this behaviour.

Figure 5.5 shows some structural details of the complexes of 18-crown-6 with ammonium, methylammonium and adamantylammonium. First, it has to be noted that the distances of major H-bonds decrease with the increasing bulkiness of the substituents. Quite interestingly, each system attempts to alleviate steric repulsions not by distancing the ammonium substituent from the ligand's framework, but by rotating the ammonium cation with respect to the macrocycle. In fact, the distance between the nitrogen atom of the ammonium cation from the mean plane of the six ethereal oxygen atoms does not change significantly with the bulkiness of the substituent (Figures 5.5(a), 5.5(c), and 5.5(e)). On the other hand, the twist angle between the triangle of the three ammonium hydrogens and the triangle obtained by linking the three ethereal oxygens closest to NH_4^+ increases with the increasing bulkiness of the substituent (H, 0° < Me, 5° < *tert*-But, 15°). Rotation moves the hydrogens far from interacting oxygens: the greater the angle, the longer the N–H$\cdots$O distance and the

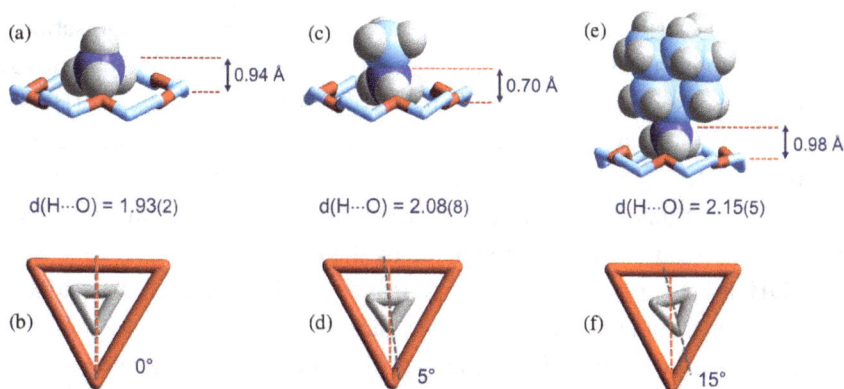

Figure 5.5. H-bond interactions and steric effects in the complexation of alkylammonium cations by 18-crown-6 (L); (a), (c), (e): distances between the nitrogen atom of ammonium and the mean plane of the six ethereal oxygen atoms of the crown; (b), (d), (f): triangles (nearly equilateral) obtained by linking the three H-bonded hydrogens of the ammonium ion and the three closest oxygens of the crown, with the angle resulting from the intersection of the triangle of the hydrogens and of the triangle of the oxygens (twist angle). (a) [(NH$_4$)(L)]NCS,[46] (c) [(MeNH$_3$)(L)]CF$_3$SO$_3$,[53] (e) [(adamantylammonium)(L)]½ZnCl$_4$[54]; counterions and C–H hydrogen atoms have been omitted for clarity.

weaker the H-bond interaction (and, as a consequence, the lower the log K value of the association equilibrium).

Ammonia is a close relative of water (the sister?). From this point of view, ammonium should be the cousin of oxonium. Figure 5.6 illustrates some structural details of the isoelectronic species NH$_3$, H$_2$O, NH$_4^+$ and H$_3$O$^+$ in the gas phase.

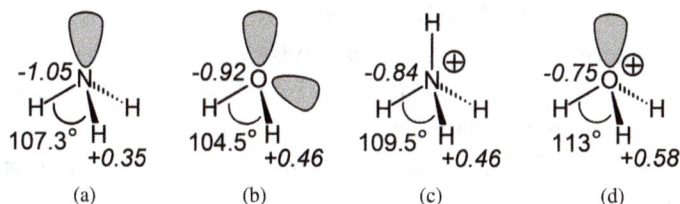

Figure 5.6. Structural details of ammonia (a) and water (b) molecules and ammonium (c) and oxonium (d) cations: HXH bond angles (X = N, O) and atomic partial charges (calculated through *Natural Population Analysis*). All species are isoelectronic.

In all cases, the central atom exhibits sp^3 (tetrahedral) hybridisation. While ammonia and water deviate from regular tetrahedral geometry due to the repulsion exerted by one (HNH angle 107.3°) and two lone pairs (HOH angle 104.5°), the symmetric ammonium ion shows a regular tetrahedral geometry (HNH angle 109.5°). In the oxonium cation, the reciprocal electrostatic repulsions exerted by the hydrogen atoms, which detain a partial positive charge, overcome the repulsion by the lone pair. As a consequence, the HOH angle is not contracted, but expanded to 113°, and H_3O^+ assumes a flattened pyramidal structure. Atomic partial charges, calculated through *Natural Population Analysis*, on the hydrogen atoms suggest that oxonium (+0.58) is a stronger H-bond donor than ammonium (+0.46). According to the Brønsted theory, H_3O^+ is the acid of limiting strength in an aqueous solution ($pK_A = 0$, every acid of higher strength transfers a proton to a water molecule making H_3O^+), whereas ammonium is a weak acid ($pK_A = 9.25$).

Analogous with ammonium, oxonium gives solid salts, e.g. $[H_3O]ClO_4$, oxonium perchlorate (often erroneously written as $HClO_4 \cdot H_2O$, hydrated perchloric acid, but nobody would write $HClO_4 \cdot NH_3$ instead of NH_4ClO_4!). However, due to its high acidity, oxonium tends to transfer a proton to the counteranion and forms salts only with anions of extremely low basicity: with ClO_4^- ($pK_A = -10$) for instance, but not with Cl^- ($pK_A = -8$). On the other hand, the much less acidic ammonium gives stable salts with most inorganic and organic anions.

Figure 5.7(a) shows the crystal structure of the complex of oxonium with 18-crown-6.[55] In analogy with the corresponding ammonium complex, each hydrogen atom of H_3O^+ establishes (i) a major H-bond interaction with the facing oxygen atom and (ii) two minor interactions with the two oxygen atoms adjacent to the facing one. The distances of the three major bonds (1.69, 1.69, 1.74 Å) are those typically observed for an O–H⋯O pattern, distinctly lower than those observed in the corresponding ammonium complex (1.91, 1.91, 1.95 Å). Such a situation is achieved (i) through the complete flattening of the OH_3 pyramid, which is revealed by value of the HOH angles, expanded from 113° of the free oxonion to the nearly 120° of the complexed cation and (ii) by the moving of the H_3O^+ cation towards the O6 cavity: in fact, the oxonium oxygen atom is only 0.04 Å above the mean plane of the six ethereal oxygen atoms (see Figure 5.7(b)), to be compared to the 0.94 Å of the N---O6 distance of the corresponding ammonium complex. Indeed, symmetry and coplanarity,

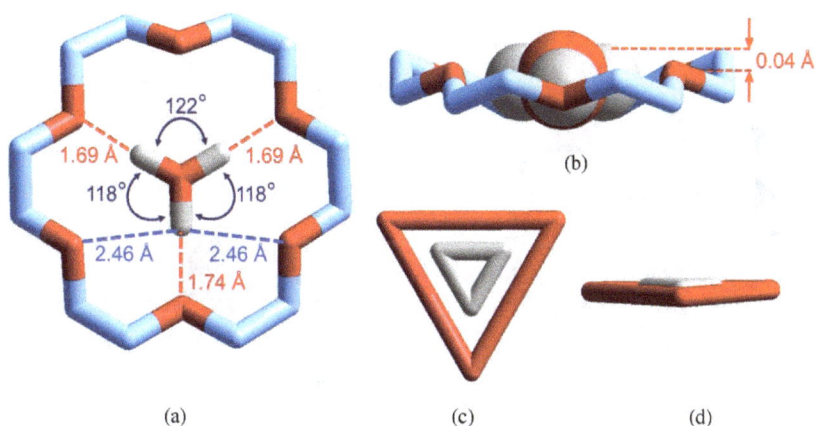

Figure 5.7. The structure of {[H$_3$O](18-crown)}$_2$ (Mo^{VI}O$_4$) salt[55]; C–H hydrogen atoms and molybdate counterion have been omitted for clarity; (a) top view and structural details of the oxonium–crown complex; (b) side view of the complex; (c) and (d) triangles obtained by linking the hydrogens of H$_3$O$^+$ (white) and the three closest ethereal oxygens of the crown (red).

illustrated in Figures 5.7(c) and 5.7(d), ensure the formation of an especially stable H-bond complex. Unfortunately, studies on the thermodynamic stability of the {[H$_3$O](18-crown)}$^+$ complex have not been carried out yet.

The coordination chemistry of oxonium cannot be extended to cryptands because the strong acid H$_3$O$^+$ would transfer its proton to one of the bridgehead tertiary nitrogen atoms. Thus, it happens that on evaporation of an acidic solution of 2.2.2-crypt, the diprotonated cryptand crystallises in a form containing in its cavity the conjugate base of H$_3$O$^+$:H$_2$O. An example is shown in Figure 5.8, which displays the crystal structure of the salt oxonium dihydrogen-(2.2.2-crypt) tribromide hydrate.[56] It is observed that the diprotonated cryptand includes a water molecule, while an oxonium cation is well distanced in the crystal from the receptor.

In Figure 5.8(a), dashed lines indicate the four hydrogen bonds formed by the included water molecule and the diprotonated cryptand: two O–H$\cdots$O$_{ethereal}$ (red lines) and two N$\cdots$H–O$_{water}$ (blue lines). Figure 5.8(b) shows that each one of the three O–H fragments of oxonium interacts with one bromide ion. In fact, Br$^-$ is not basic enough to uptake a proton from H$_3$O$^+$.

Here ends the short happy life of the oxonium cation in coordination chemistry.

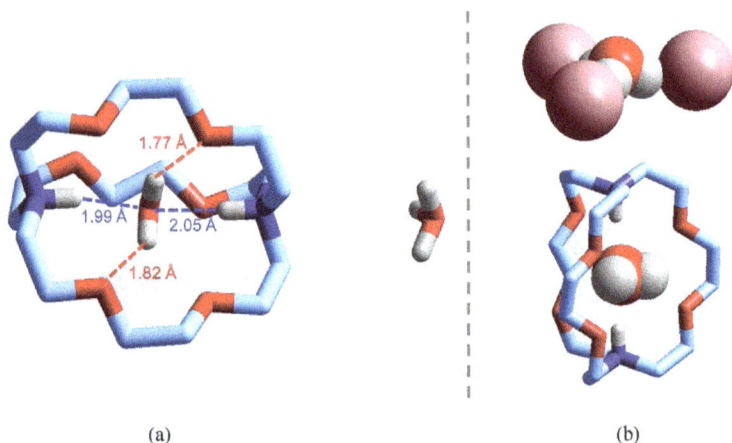

(a)　　　　　　　　　　　　　　　　　　　　(b)

Figure 5.8. The crystal structure of the salt oxonium dihydrogen-(2.2.2-crypt) tribromide hydrate[56]: (a) dashed lines indicate the H-bonds established by the included water molecule and the diprotonated cryptand: two $O–H\cdots O_{ethereal}$ (red lines) and two $N\cdots H–O_{water}$ (blue lines); H_3O^+ is placed far away from the cryptand; (b) the same structure including bromide anions: each Br^- interacts with one hydrogen of H_3O^+.

The divine origin of the -onium desinence

We cannot say what came first, the chicken or the egg. But we know that ammonium came before ammonia. The story begins during 21st century BC, when Amun (or Amon; Ancient Greek: Αμμων, Hámmōn), King of Thebes, Egypt, rose to the position of patron deity of the town. He was later fused with the Sun god, Ra. In a few centuries, Amun Ra reached a dominant position in the Egyptian pantheon to finally hold the rank of King of Gods. Then, he came to be worshipped outside of Egypt, in Ancient Libya and Nubia, while in Ancient Greece he was identified with Zeus (and named Zeus Ammon). Temples devoted to Ammon were built in Nubia, Sudan and Lybia. One of the most famous was that in the Siwa Oasis, Lybia, which hosted a great Oracle, consulted, among others, by Alexander the Great and by Hannibal. Worshippers typically reached the Siwa Oasis and the Temple of the Oracle riding camels, which were 'parked' near the shrine. At the end of the day, the camels left tons of dung on the sand. For hygiene reasons, the manure was burnt, leaving on the sand ashes containing an inorganic salt, NH_4Cl, which was called salt of Ammon, in Latin 'sal ammoniacus'. *Sal-ammoniac* had a prominent role in alchemy and boasted its own alchemical symbol: an asterisk. It shared with other salts, e.g. marine salt (NaCl) and saltpeter (KNO$_3$), solubility in water, but showed an outstanding difference: while other salts melted on heating, sal-ammoniac sublimated. First-year students studying Le Chatelier's principle learn that the apparent sublimation is based on the following temperature-dependent equilibrium:

$$NH_4Cl(s) + heat \leftrightarrows NH_3(g) + HCl(g)$$

On heating, solid NH$_4$Cl decomposes to give gaseous NH$_3$ and HCl (the equilibrium is displaced to the right). However, if gaseous NH$_3$ and HCl, diffusing in the air, find the cold surface of a watch glass, they recombine to give solid NH$_4$Cl, which crystallises on the glass (equilibrium displaced to the left). Indeed, 'sublimation' was the procedure used by Romans to recover sal ammoniac from the ashes near the Ammon Temple at Siwa Oasis. In the 15th century, Basilius Valentinus, a German Benedictine monk and an alchemist, showed that ammonia could be obtained by the action of alkali on sal-ammoniac. Later, ammonium granted the suffix -onium to the cations derived by addition of a proton to a mononuclear hydride of the XV, XVI and XVII groups (e.g. PH$_4^+$ phosphonium, H$_3$S$^+$, sulfonium, H$_2$Cl$^+$ chloronium).

6

The Age of Anion Coordination Chemistry

Supramolecular chemistry is the chemistry of non-covalent interactions.[57] Non-covalent interactions are typically (but not necessarily) weak and are (necessarily) quickly reversible. They include: hydrogen bonding, electrostatic interactions, π–π donor–acceptor interactions, metal–ligand interactions. The presence of metal coordination chemistry among the sub-topics of supramolecular chemistry creates a temporal paradox. In fact, coordination chemistry was officially recognised in 1893, with the publication of the monumental paper of Alfred Werner on the amine complexes of substitutionally inert transition metal ions,[58] and was universally recognised with the assignment to Werner of the Nobel Prize in 1913.

While (i) the existence of intermolecular forces was first postulated by van der Waals in 1873,[59] and (ii) Latimer and Rodebush described the formation of hydrogen bonds in water in 1920,[60] supramolecular chemistry was established as a well-defined discipline during the 60s. Thus, it happens that supramolecular chemistry (the 'mother') is much younger than metal coordination chemistry (her 'first daughter'). More or less at the same time, during the 60s, the 'little sister' of metal coordination chemistry was born: anion coordination chemistry, which completed an intricate family network.

Anion coordination chemistry is concerned with the formation of complexes constituted by an anion and by its receptor. The receptor is a concave molecular system containing one or more points of interaction — positive charges, H-bond donors — strategically positioned inside the cavity. Park and Simmons in 1968 studied through ^{1}H NMR titration experiments in a 50/50 trifluoroacetic acid/water solution the complexation of chloride by doubly protonated triple-stranded diammonium alkanes of the type shown in Figure 6.1: **20**.[61]

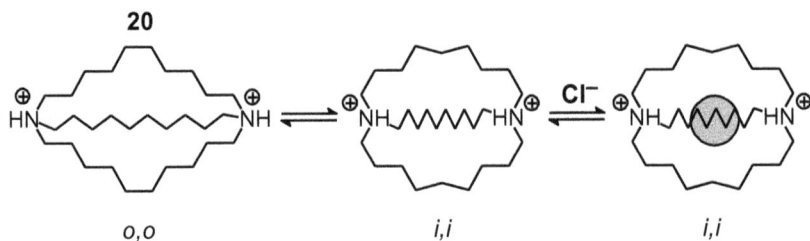

20

o,o   i,i   i,i

Figure 6.1. Chloride encapsulation by a triple-stranded diammonium alkane, **20**, in a 50% trifluoroacetic acid aqueous solution. Anion inclusion is preceded by a structural rearrangement of the diammonium receptor from the *out,out* isomer (*o,o* — the two N–H fragments pointing outside of the cavity) to the *in,in* isomer (*i,i* — N–H fragments pointing inside). A chloride *katapinate* complex is formed.[61]

They preliminarily observed that the diammonium ion **20** exists as an equilibrium mixture of the *out,out* (*o,o*) isomer, in which the ammonium N–H fragments point outside the cavity (23%), and of the *in,in* (*i,i*) isomer, with N–H fragments pointing inside the cavity (77%). The interconversion equilibrium is fast. Quite interestingly, progressive addition of NaCl made the resonance of –CH$_2$– protons shift upfield, which indicated the occurrence of a fast anion encapsulation process. The constant K for the inclusion equilibrium was estimated to be greater than 10. A similar behaviour was observed on addition of Br$^-$ and I$^-$, which demonstrated the poor selectivity of receptor **20** towards halides, a feature probably related to the flexibility of the aliphatic chains. No crystal structures have been determined for **20** and its halide complexes. The calculated structures in Figure 6.2 provide tentative details of the encapsulation process.

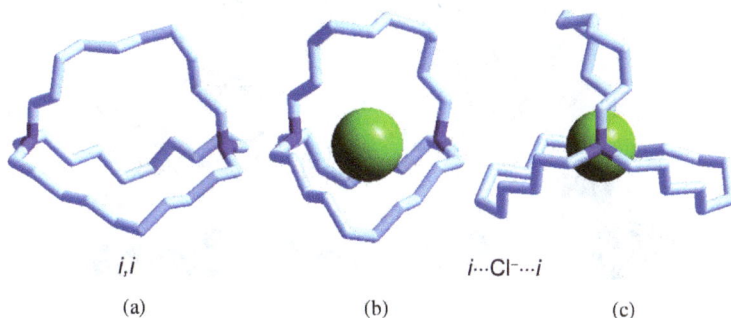

i,i   *i⋯Cl⁻⋯i*

(a)   (b)   (c)

Figure 6.2. The calculated structures: (a) of the *in,in* isomer of the diammonium receptor **20**, and ((b) and (c)) of its chloride inclusion complex, lateral and longitudinal views. Hydrogen atoms have been omitted for clarity. The crystal structure of the complex salt $[H_{13}O_6][L\cdots Cl]Cl_2$ has been determined,[62] but unavailability of coordinates at the Cambridge Crystallographic Database prevented any redrawing of such a structure. Interest in the paper was also motivated by the presence of the cation $H_{13}O_6^+$, an unusual form of the hydrated proton, displaying an ordered network of short and symmetric hydrogen bonds.

In particular, it is observed that in the *i,i* isomer the receptor (Figure 6.2(a)) has significantly contracted its cavity in order to accommodate the anion and to establish collinear hydrogen bonding interactions N–H⋯Cl⋯H–N (Figures 6.2(b) and 6.2(c)).

The authors gave to the above-described anion inclusion complexes the name of 'katapinate', from the Ancient Greek verb καταπίνειν, to swallow, thus considering the receptor as a big fish swimming in solution and looking for a smaller fish (the anion) to ingest (see the cartoon in Figure 6.3). Later, the supramolecular community, following the nomenclature introduced by Lehn with cryptands, named the triple-strand diammonium receptors of Park and Simmons *katapinands*.[63]

Park and Simmons worked at Du Pont de Nemours Company, in Wilmington, Delaware, and it is probable that a different compelling industrial project diverted them from this promising area of research.

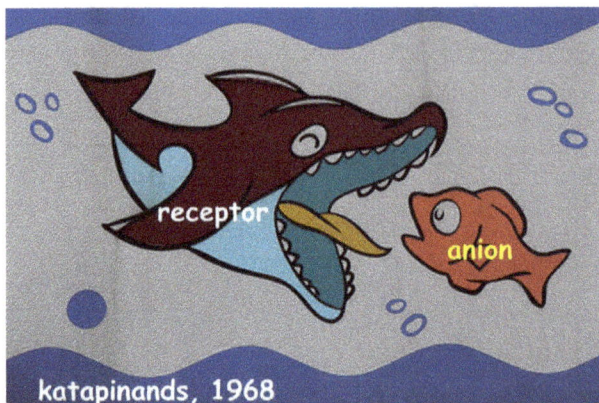

Figure 6.3. The origin of the term 'katapinate'[61]: from the Ancient Greek verb καταπίνειν, to swallow. The cartoon illustrates the metaphor: a big fish (the receptor) swallows a small fish (the anion).

Thus, they did not continue in this vein, and the design of anion receptors was abandoned for eight years, until Lehn realised that protonation of the four amine groups of the spherical cryptand **19** would give a molecular system containing four positive charges placed at the corners of a tetrahedron.[64] The tetrahedron is the ideal arrangement for the H-bond interaction with a 'spherical' anion, i.e. a monoatomic species with a noble gas electronic configuration, which, in fact, points its lone pairs towards the corners of a tetrahedron. (Figure 6.4).

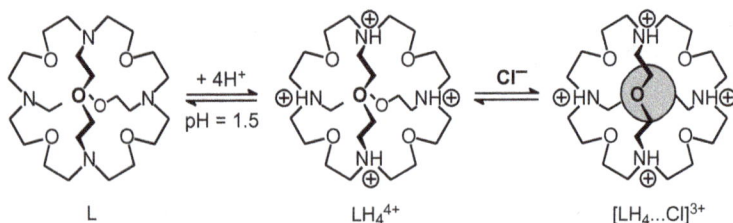

Figure 6.4. Consecutive equilibria for protonation of the spherical cryptand **19** (L) and inclusion of Cl^- in the tetraprotonated receptor LH_4^{4+}.[64] In the $[LH_4 \cdots Cl]^{3+}$ complex, the anion is tetrahedrally coordinated by four N–H fragments, as shown in the crystal structure.[52] Unavailability of coordinates at the Cambridge Crystallographic Database prevented any redrawing of such a structure.

Equilibrium studies showed that, at pH = 1.5, molecule **19** (L) exists in a tetraprotonated form, LH_4^{4+}. On addition of Me_4NCl to a solution of **19** adjusted to pH 1.5 with nitric acid, the $[LH_4 \cdots Cl]^{3+}$ complex forms, for which a log $K > 4$ was estimated for the formation equilibrium. The crystal structure of $[LH_4 \cdots Cl]Cl_3 \cdot 7H_2O$ was determined, which showed that the chloride ion lies at the centre of the intramolecular cavity and receives four H-bonds from the four N–H fragments, according to an N–H$\cdots$N linear arrangement. LH_4^{4+} shows a much lower affinity for Br⁻ ($K < 10$) and does not include I⁻ at all, demonstrating a pronounced size selectivity in favour of chloride. It has to be noted that inclusion of chloride in the tailor-made tetraprotonated spherical cryptand **19** is favoured by more than three orders of magnitude with respect to the inclusion of the diprotonated katapinand **20**. In this regard, it has to be observed that complexation by the katapinand involves the formation of only two H-bonds and a dramatically endergonic conformational rearrangement of the receptor framework.

The tertiary ammonium group, R_3NH^+, is an excellent point of interaction for anions as it provides an attractive electrostatic field and donates one hydrogen bond. However, maintenance of these features requires that investigations are carried out in a protic medium and in a definitely acidic region. On the other hand, a permanent positive charge can be imparted to a receptor by alkylating, rather than protonating, the tertiary amine groups. Such a progress in anion receptor design was achieved by Franz Schmidtchen, Technical University of Munich, who, in 1977, reported the synthesis of receptor **21**, which contains four quaternary nitrogen atoms linked by $-(CH_2)_6-$ aliphatic chains.[65] In particular, each nitrogen is linked to the other three nitrogen atoms in such a manner which is expected to generate a tetrahedral arrangement of the four quaternary ammonium groups.

The crystal structure of the iodide inclusion complex is shown in Figure 6.5. I⁻ is placed inside the cavity, and its position is coincident with that of the centroid of the N4 tetrahedron. The complex is held together only by coulombic interactions. In particular, the tetrahedral arrangement of the quaternary ammonium groups maximises attractive electrostatic interactions with the anion on one side, and minimises electrostatic repulsions between positive charges on the other. The tetrapositive receptor

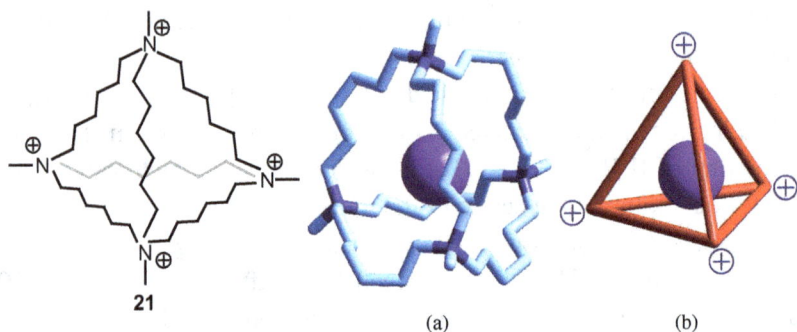

Figure 6.5. Anion inclusion in the macrotricyclic quaternary ammonium receptor **21** (L^{4+}): (a) crystal structure of complex salt $[L \cdots I]I_3 \cdot H_2O \cdot MeCN$; (b) the iodide ion is placed in the centre of the nearly regular tetrahedron described by the four nitrogen atoms: tetrahedron side 7.4 ± 0.3 Å, tetrahedron angle $57 \pm 1°$. The $N \cdots I$ distance is 4.54 ± 0.04 Å.[65]

forms stable complexes in water with Br^- and I^- (log K = 3.0 and 2.7, respectively). Increasing the length of the linking aliphatic chains from $-(CH_2)_6-$ to $-(CH_2)_8-$ inverts selectivity in favour of I^- (log K = 2.0 for Br^- and 2.5 for I^-).[65]

The three above-mentioned papers (Park and Simmons, Lehn, Schimdtchen) provided the foundation for a lively branch of supramolecular chemistry: the so-called 'anion coordination chemistry'. This name raises a point of discussion: the term 'coordination' was introduced by Alfred Werner and refers to an individual (the metal ion) which puts in order around itself several objects (ammonia molecules), usually four or six (the coordination number). On the contrary, the anion has only one object to interact with (the receptor), with no problems of coordination number and geometrical arrangement. In particular, a given anion X^- is not able to put in order around itself a number (four or six) of unidentate molecules, as a metal ion does (e.g. with six ammonia molecules). This is essentially due to the poor intensity of anion–receptor interaction (whether electrostatic or H-bond or both). However, the two classes of compounds share a feature of the interaction not present in covalent bond: a quick reversibility, typical of the 'coordinate' bond. This may account for the broadly accepted term for this topic of 'anion coordination chemistry'.

Finally, it has to be mentioned that the first studies on anion coordination chemistry were carried out by Park and Simmons in 1968 in the laboratories of DuPont at Wilmington, the same place where, one year before, Pedersen had completed his seminal work on crown ethers and their alkali metal complexes. Indeed, the birth of supramolecular chemistry is and will be permanently associated with the Laboratories of DuPont Company and to the ingenuity and skilfulness of its researchers.

7

The Appearance of
Bistren Cryptands

In 1977, Lehn synthesised a new macrobicyclic system with a molecular framework similar to that of the classical cryptands, with two bridgehead tertiary amine nitrogen atoms (**22**).[66] Each linking chain contained a nitrogen, an oxygen and another nitrogen, separated by $-CH_2CH_2-$ spacers.

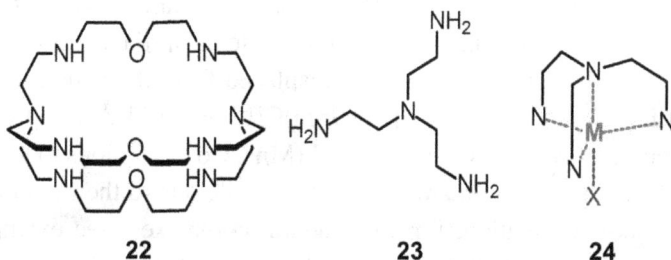

| 22 | 23 | 24 |

The new cryptand was named 'bistren', considering that it can be formally obtained by linking with $-CH_2CH_2OCH_2CH_2-$ spacers the primary amine nitrogen atoms of two facing *tren* subunits. Tren (**23**, tris (2-aminoethyl)amine) is the archetypal tripodal ligand of metal coordination chemistry. Due to the branched connectivity of the four amine nitrogen atoms, it favours five-coordination: the four amine groups are positioned at four corners of a trigonal bipyramid, while an exotic unidentate ligand X (either a molecule or an anion) occupies the vacant axial site (**24**). A metal ion showing a special affinity for tren and five-coordination is copper(II). Figure 7.1 shows the crystal structures of the two complexes [Cu^{II}(tren)Cl]$^+$ (Figure 7.1(a))[67] and [Cu^{II}(tren)N$_3$]$^+$ (Figure 7.1(b)).[68]

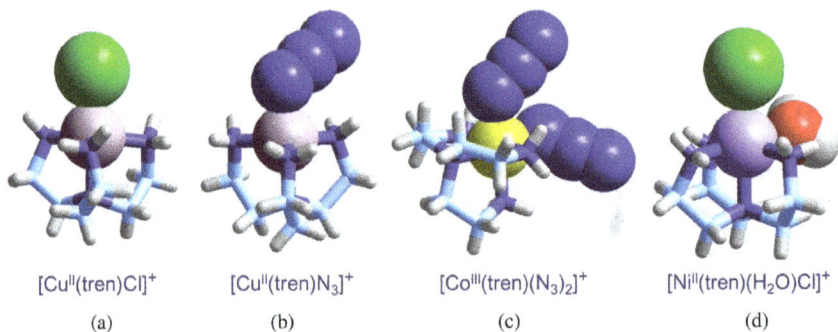

| [Cu^{II}(tren)Cl]$^+$ | [Cu^{II}(tren)N$_3$]$^+$ | [Co^{III}(tren)(N$_3$)$_2$]$^+$ | [Ni^{II}(tren)(H$_2$O)Cl]$^+$ |
| (a) | (b) | (c) | (d) |

Figure 7.1. The crystal structures of the complex salts (a) [Cu^{II}(tren)Cl]BPh$_4$,[67] (b) [Cu^{II}(tren)N$_3$]ClO$_4$,[68] (c) [Co^{III}(tren)(N$_3$)$_2$]ClO$_4$,[69] (d) [Ni^{II}(tren)(H$_2$O)Cl]Cl·H$_2$O.[70] Counteranions and solvating molecules have been omitted for clarity. The two Cu^{II} complexes are five-coordinate and show a trigonal bipyramidal geometry; the Co^{III} and Ni^{II} complexes are six-coordinate with an octahedral geometry.

Both complexes exhibit a regular trigonal bipyramidal geometry, with the anion placed in the axial position opposite to the tertiary amine nitrogen atom of tren. The Cu^{II} ion is slightly displaced from the equatorial plane towards the anion (0.24 Å for [Cu^{II}(tren)Cl]$^+$ and 0.23 Å for [Cu^{II}(tren) N$_3$]$^+$). Tren interacts with other 3d metal (Mn^{II}, Co^{II}, Zn^{II}) ions to give five-coordinate complexes with anions. Co^{III} and Ni^{II}, due to their natural penchant for six-coordination, form octahedral complexes (see examples in Figures 7.1(c)[69] and 7.1(d),[70] uptaking a further unidentate ligand, either an anion or a solvent molecule. In view of lability and high position of copper(II) in the *spectrochemical series* of metal ions, [Cu^{II}(tren)]$^{2+}$ represents an ideal subunit for exchanging and binding anions.

On these premises, Lehn suggested that the bistren cryptand **22** could undergo the two-step process, illustrated in Figure 7.2(a).

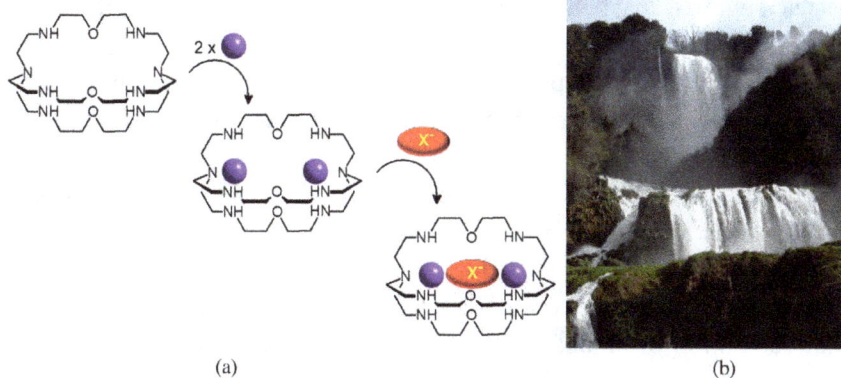

Figure 7.2. (a) Two-step cascade process involving (i) the complexation of two Cu^{II} ions by bistren, (ii) the inclusion of an ambidentate anion X^-, whose donor atom(s) fill(s) the vacant axial sites of each metal centre[66]; (b) Marmore's Cascade Falls, a man-made water-fall, which was created by Romans in 271 BC to divert stagnant water. It is located near Terni, a provincial capital in the Italian region of Umbria. Its total height is 165 m, which makes it the highest artificial waterfall in Europe. It consists of three steps, and the top one is the tallest, 83 m.

In the first step, two Cu^{II} ions enter the cavity of the cryptand, and each one establishes coordinative interactions with the four nitrogen atoms of one tren subunit, thus remaining coordinatively unsaturated. In the second step, an ambidentate anion, either monoatomic (e.g. Cl^-) or polyatomic (e.g. N_3^-), goes to bridge the two metal centres, which reach coordinative saturation. Lehn proposed for ternary complexes of type $[Cu^{II}_2(bistren)X]^{3+}$ (X^- = anion) the name *cascade* complexes,[66] consider-ing that their formation is a sequence of enchained spontaneous processes, like those observed in a series of small waterfalls, falling in stages down a steep rocky slope (see Figure 7.2(b) for a beautiful example). In particu-lar, (i) first, the ligand chooses the metal ions, (ii) second, the metal ions choose the ambidentate anion better satisfying their electronic and geo-metrical requirements. While the mononuclear $[Cu^{II}(tren)]^{2+}$ complex chooses the anion capable of establishing the strongest coordinative interac-tions (the highest in the spectrochemical series), the choice by the dinuclear complex $[Cu^{II}_2(bistren)]^{4+}$ is primarily based on geometrical features: the selected anion is that better encompassing with its donor atom(s) the inter-metallic distance.

The equilibria of stepwise complexation of two Cu^{II} ions by bistren **22** were investigated through potentiometric titration experiments in an aqueous solution 1 M in $NaClO_4$ at 25°C.[71] Then, titration experiments were carried out in solutions 1 M both in $NaClO_4$ and in NaX (X = F, Cl, Br, I), which allowed to evaluate log K values for the $[Cu^{II}_2(bistren)]^{4+}$ + X^- ⇆ $[Cu^{II}_2(bistren)X]^{3+}$ inclusion equilibria.[72] This study demonstrated that dimetallic cascade complexes could represent effective systems for the recognition of anions in water, operating on the basis of both geo-metrical and bonding features.

However, the bistren cryptand **22** can take another way to tempt ani-ons into entering its cavity: that of protonating its amine nitrogen atoms.[73] On potentiometric titration studies over the 2–12 pH interval, six pK_a values could be evaluated, ranging from 10.0 to 5.7 (at 25°C) and corre-sponding to the protonation of the six secondary amine groups of **22**.[74] The protonation of the two bridgehead tertiary amine groups of **22**, due to the strong electrostatic repulsions, could take place only in very acidic solutions (pH < 2). Figure 7.3 shows the distribution curves for the species present at the equilibrium over the 2–12 pH interval.

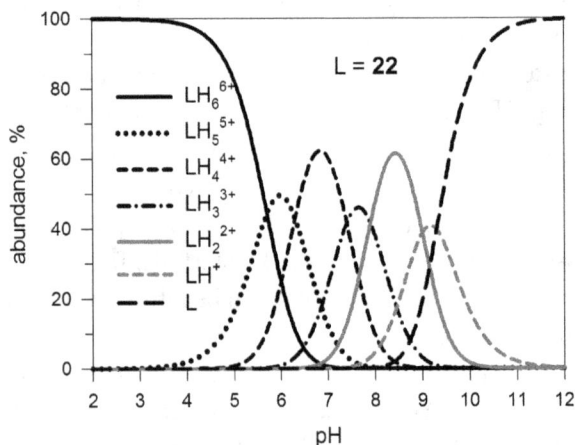

Figure 7.3. Concentration profiles of a 10^{-4} M bistren cryptand **22** (L) solution over the pH range 2–12. Curves were calculated on the basis of stepwise protonation constants from Ref. 74. Abundance is given by $([LH_n^{n+}]/C_L) \times 100$ (C_L, analytical concentration of L).

Noticeably, the LH_6^{6+} cation is the only species present below pH 4 and provides a system of six positive charges symmetrically positioned within an ellipsoidal cavity, suitable for encapsulation of anions. Thus, a second type of cascade process can be designed for the inclusion of anions, illustrated in Figure 7.4.

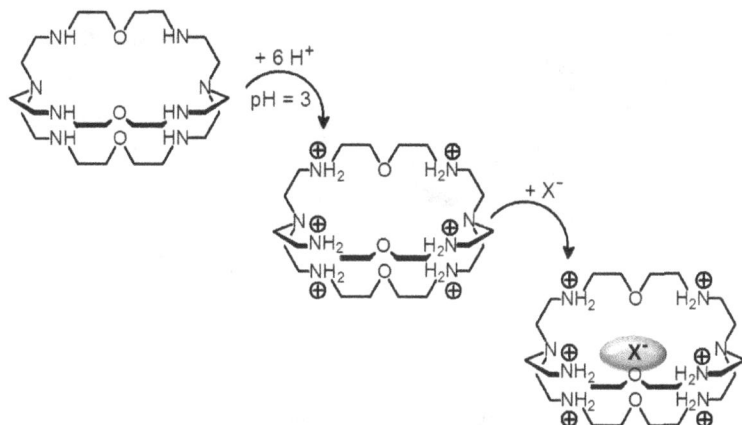

Figure 7.4. Cascade process for the inclusion of an anion X^- into the hexaprotonated bistren cryptand **22**.[73] In the 2–4 pH interval, only the LH_6^{6+} is present in solution, ready to incorporate anions of complementary size and shape.

On titration of a solution of the bistren cryptand **22** with NaN_3, adjusted to pH 5 with perchloric acid, ^{13}C resonance shifts progressively increased to reach a plateau after the addition of 1 equivalent of azide, indicating the formation of a 1:1 complex.[73] The distribution diagram in Figure 7.1 shows that at pH 5 the bistren receptor (= L) is present at ca. 80% as LH_6^{6+} and at ca. 20% as LH_5^{5+}. From the titration profiles (^{13}C NMR shift against anion equivalent), a conditional association constant log $K = 4.3 \pm 0.3$ was evaluated.[73] On titration with a variety of other mono- and poly-atomic anions (including halides), log K values lower by at least two orders of magnitude were determined. Noticeably, titration of a solution of tren with azide at pH = 5, containing the triprotonated species $trenH_3^{3+}$, did not cause any change in the ^{13}C spectra, thus excluding the formation of any receptor–anion complex. This demonstrated the existence of a cryptate effect in anion coordination chemistry also. The formation of an inclusion complex with the formula $[LH_6\cdots N_3]^{5+}$ was confirmed later by X-ray diffraction studies on the crystalline salt $[LH_6\cdots N_3](Cl)(PF_6)_4\cdot 3H_2O$, whose structure is shown in Figure 7.5(a).[75]

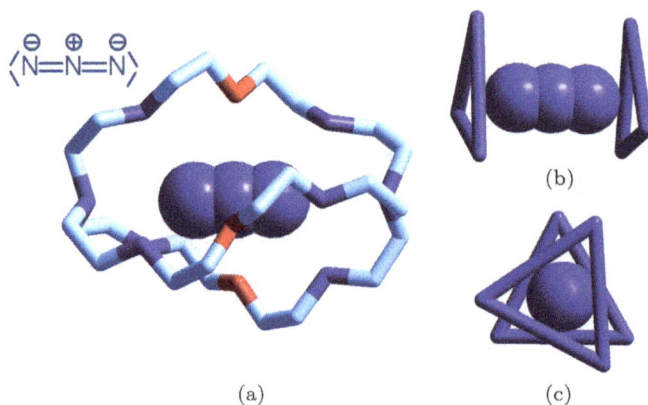

Figure 7.5.   (a) The crystal structure of the complex salt $[LH_6\cdots N_3](Cl)(PF_6)_4\cdot 3H_2O$ (L = **22**).[75] Counteranions and solvational waters have been omitted for clarity. Each terminal azide nitrogen atom receives three hydrogen bonds from the secondary ammonium groups of the proximate tren subunit; (b) nitrogen atoms of the H-bond donating N–H fragments of each tren subunit have been linked to give a triangle, lateral view; (c) top view, showing that the linear N_3^- anion lies on the ternary axis of the cryptate complex.

Each terminal atom of the azide ion, which, according to the Lewis structural formula shown in Figure 7.5 detains a formally negative charge, receives three equivalent hydrogen bonds from the three secondary ammonium groups of the close tren subunit. In particular the N(azide)---N(ammonium) distances range from 2.97 to 3.00 Å. For better illustrating the geometrical features of the anion–receptor complex, the protonated nitrogen atoms of each tren subunit have been linked together, to give a triangle (Figure 7.5(b)). The three ammonium groups at the three vertices of each triangle interact with one terminal nitrogen atom of the azide ion. The torsional angle formed by the two triangles, θ, is $27 \pm 1°$, a value intermediate between 0°, corresponding to an *eclipsed*, and 60°, a *staggered* arrangement of the tren subunits (see Figure 7.5(c)). The linear N_3^- anion lies along the ternary axis of the H-bond complex (the line joining the bridgehead ternary amine nitrogen atoms of the bistren cryptand). The LH_6^{6+} receptor seems to offer to N_3^- a tailor-made cavity.

Figure 7.6(a) shows the crystal structure of the $[LH_6^{6+}\cdots Br]Br_5$ complex salt.[75]

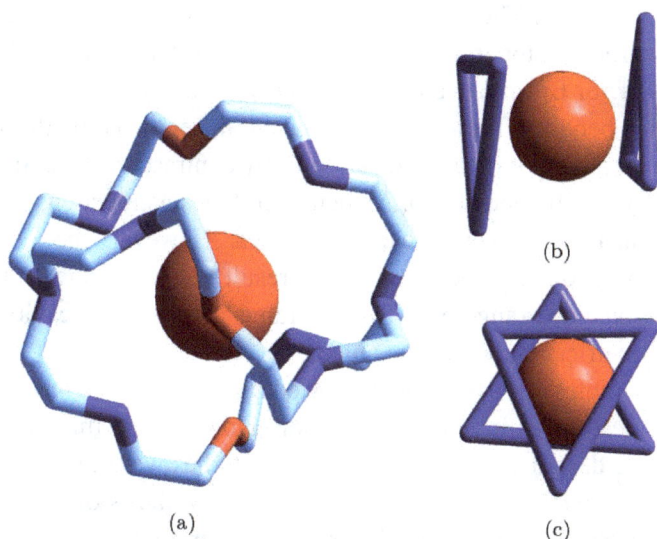

Figure 7.6. (a) The crystal structure of the complex salt $[LH_6\cdots Br]Br_5$ (L = **22**).[75] Counteranions and hydrogen atoms have been omitted for clarity. The encapsulated bromide anion receives six hydrogen bonds from the secondary ammonium of the receptor; (b) H-bond donating nitrogen atoms of each tren subunit have been linked to give a triangle, lateral view; (c) top view: twist angle $\theta = 58°$.

The bromide ion is well included in the receptor's cavity and receives six nearly equivalent hydrogen bonds from the six secondary ammonium group: five of the $N(ammonium)\cdots Br$ distances range from 3.32 to 3.37 Å, one is 3.45 Å. The 'length' of the cryptand, i.e. the distance between the bridgehead nitrogen atoms, N_{tert}---N_{tert}, is distinctly shorter than that observed for the corresponding azide complex (7.53 vs 8.84 Å), which indicates that the receptor has significantly contracted its cavity in order to provide a spheroidal cavity and accommodate the monoatomic anion. Thus, thanks to the flexibility of the $-(CH_2)_2O(CH_2)_2-$ spacers, the receptor can offer both a spheroidal and an ellipsoidal cavity, depending upon the geometrical requirements of the anion. It has to be observed that the 'length' of the cage is modulated by a 'spring and screw' mechanism, which operates through the rotation of one of the two triangles described by the secondary ammonium nitrogen atoms of each tren subunit with respect to the other: the more pronounced the rotation, the shorter the N_{tert}---N_{tert} distance. Thus, generation of the ellipsoidal cavity for N_3^- is achieved through a torsion of 27°, whereas for the formation of the spheroidal cavity for Br$^-$, the two triangles must be rotated by 58°. As a result, the Br$^-$ ion benefits from a nearly octahedral geometry of the H-bond network (for a regular octahedron $\theta = 60°$; by contrast, $\theta = 0°$ corresponds a trigonal prismatic geometry). An octahedral arrangement ensures minimised repulsions between positive charges. However, the octahedral coordination geometry seems to be the result of the 'spring and screw' conformational rearrangement for providing a spheroidal cavity, rather than a geometrical choice of the anion.

The examples discussed above showed the potential and the versatility of bistren derivatives in anion recognition. However, no further developments in the design were observed, and so studies were restricted to the bistren prototype **22**. The reason is simple: the synthesis of **22** is lengthy and complicated and may have discouraged chemists with a mainly inorganic background, naturally interested in the coordination chemistry.[73] The complexity of the synthesis of **22** is illustrated in Figure 7.7.

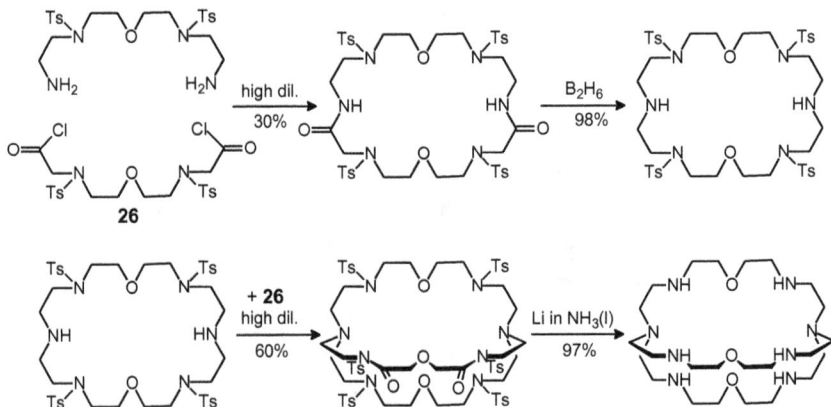

Figure 7.7. The multistep synthesis of bistren cryptand **22**. The process involves two high-dilution reactions and is characterised by an overall yield of 17%.[73]

The synthesis includes two high-dilution processes and has an overall yield of 17%. In addition, intermediates for cyclisation processes illustrated in Figure 7.7 are not commercial, but require specific syntheses, as sketched in Figure 7.8.

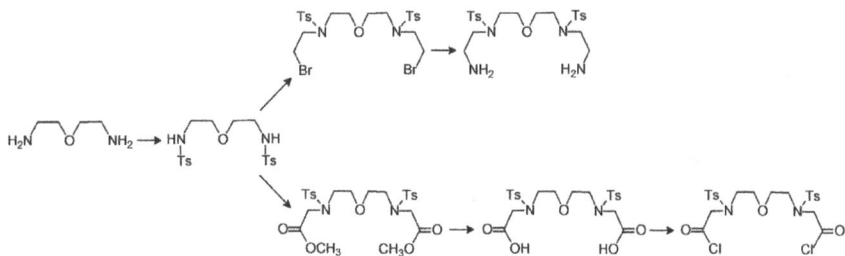

Figure 7.8. The syntheses of intermediates for the reactions of Figure 7.7.

Things changed abruptly and drastically when a simple one-pot synthesis of bistren cryptands based on Schiff base condensation was introduced.

8

Bistren Cryptands Grow and Multiply

In 1987, Lehn and coworkers reported the direct and uncomplicated synthesis of the bistren cryptand **27**,[76] which is illustrated in Figure 8.1.

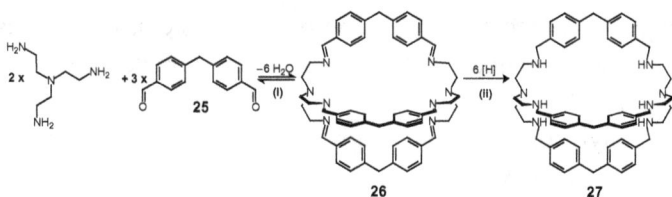

Figure 8.1. The 2 × 3 Schiff base condensation of tren and dialdehyde **25** to give the unsaturated bistren cryptand **26**; process (i) is reversible, and the six imine bonds can hydrolyse and give back the reactants. The imine bond can be easily hydrogenated to give the kinetically stable saturated cryptand **27**.[76]

The process involves the Schiff base condensation of 2 molecules of tren (providing on the whole 6 primary amine groups) and 3 molecules of dialdehyde (providing 6 carbonyl groups) to give a cage-shaped macrobicycle containing 6 imine bonds. In practice, a solution of dialdehyde **25**

(1.71 mmol) in MeCN was added dropwise to a solution of tren (1.14 mmol) in MeCN over a period of 15 min. A precipitate of almost pure macrobicyclic hexaimine (**26**) precipitated, which was recrystallised from MeCN/CH₂Cl₂ (yield 60%).[76] The crystal structure of the unsaturated bistren cryptand is shown in Figure 8.2.

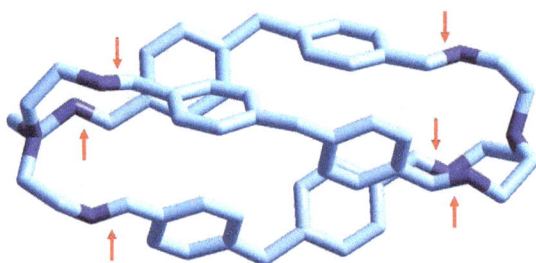

Figure 8.2. The crystal structure of the hexaimine cryptand **26**,[76] providing an elongated ellipsoidal cavity. Red arrows indicate C=N bonds. The N_{tert}---N_{tert} distance ('length' of cryptand) is 15.4 Å. Hydrogen atoms have been omitted for clarity.

The bimacrocyclic hexaimine **26** presents an elongated ellipsoidal cavity, whose major axis (the N_{tert}---N_{tert} distance) is 15.4 Å.

It may seem somewhat miraculous that 5 molecules (2 molecule of tren + 3 molecules of dialdehyde) combine to undergo Schiff base condensation and produce a complex and ordered system, which is isolated in a pure form and in good yield. This apparent miracle derives from the unique nature of the imine bond. The general mechanism of primary amine–carbonyl condensation to give an imine bond is illustrated in Figure 8.3.

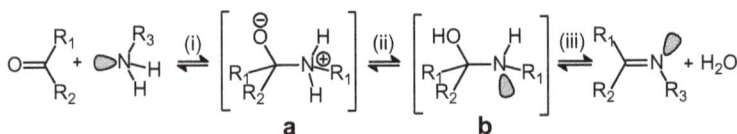

Figure 8.3. The mechanism of primary amine–carbonyl condensation to give the imine bond. All steps are quickly reversible.

In step (i), the amine nitrogen atom makes a nucleophilic addition to the carbonyl carbon atom to produce the separated charge species **a**. Then, in step (ii), a proton is transferred from the ammonium group to the negatively charged oxygen atom to give the carbinolamine intermediate **b**. Finally, **b** undergoes water elimination to give the imine derivative. Noticeably, all the species are in equilibrium, and the entire process is fast and reversible. It is in particular this fast reversibility that makes possible the synthesis of an elaborate structure. In fact, thousands and thousands of amine/carbonyl condensation processes take place according to *a trial-and-error* mechanism until the most stable product is obtained (which may have a sophisticated structure). Thanks to the speed of the equilibria, the whole synthetic process of the bistren cryptand is completed in seconds. Precipitation of the imine derivative helps to displace the equilibrium to the right. Another well-known trick for increasing yield is removal of water from the reaction medium.

The unsaturated bistren derivative **26** can coordinate only d^{10} metal ions like Cu^I and Ag^I. Figure 8.4 shows the crystal structure of the complex salt $[Cu^I_2(L)](ClO_4)_2 \cdot 2H_2O$ (L = **26**).[76]

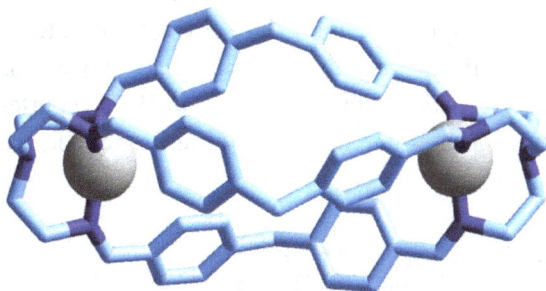

Figure 8.4. The crystal structure of the complex salt $[Cu^I_2(L)](ClO_4)_2 \cdot 2H_2O$ (L = **26**).[76] Hydrogen atoms, counteranions and solvational water molecules have been omitted for clarity. The N_{tert}---N_{tert} distance is 15.5 Å, and the Cu^I---Cu^I distance is 11.1 Å.

Each copper(I) ion interacts with one tetra-aza subunit, reaching a trigonal pyramidal four-coordination. The Cu^I–N_{imine} bonds (1.94–1.99 Å) are significantly shorter than the axial Cu^I–$N_{tert\text{-}amine}$ bond (2.25–2.26 Å). This may reflect the fact that the Cu^I–N_{imine} bond has a pronounced π character (donation from a filled $d\pi$ orbital of the metal to an empty π^* orbital of the C=N bond). The distance between the two metal centres is 11.1 Å, which would leave plenty of space for the inclusion of a polyatomic anion.

However, anions are not encouraged to enter the cavity, because Cu^I is a spherical cation and does not interact with another ligand to increase its coordination number, a tendency typical of transition metal ions. On the other hand, any attempt to complex transition metal ions (e.g. Cu^{II}) induced decomposition of the macrobicyclic ligand. In fact, Cu^{II} shows a special affinity towards the amine group (more than for an imine nitrogen atom) and tends to form a complex with the tetramine tren, thus displacing the Schiff base condensation equilibrium to the left. Decomposition is also induced by the addition of a strong acid, because H$^+$ ions go to protonate tren, another way to push back the Schiff base condensation equilibrium.

Fortunately, the 'reversible' imine bond can be 'immobilised' through hydrogenation to generate a single C–N bond. Lehn and coworkers did that with LiAlH$_4$ in THF to obtain the bistren cryptand **27**, with a 60% yield, based on the recrystallised hexahydrochloride.[76] The single C–N bond is weaker than the double C=N bond, but it is kinetically inert and guarantees permanent stability under any condition.

The convenient synthetic procedure revived the interest on bistren cryptands. A significant contribution to the topic was made by Jane Nelson (Open University, Belfast, UK) who, in 1988, reported the synthesis of a variety of octa-amine cryptands differing for the spacers linking the two tren subunits.[77]

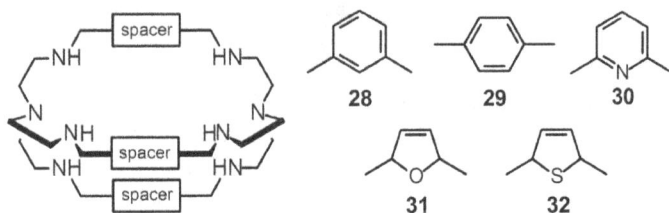

Schiff base condensation processes were carried out in refluxing EtOH with 50%–60% yield, and subsequent hydrogenation of the imine bonds was carried out with NaBH$_4$ in MeOH.

Based on our first-hand experience, 3 + 2 Schiff base condensation can be conveniently carried out in dry MeOH at room temperature on a 10^{-2} M scale. In most cases, in the presence of an aromatic dialdehyde, the bimacrocyclic hexaimine precipitates in a pure form in minutes or hours. If not, it is very probable that condensation has been unsuccessful, and any attempt to induce precipitation (concentration at rotavapor, addition of an insolubilising solvent) rarely produces any effect. As a final step, the suspension in MeOH of the hexaimine is hydrogenated with $NaBH_4$ to give the amine bistren cryptand. The availability of a variety of bistren cryptands differing in the length and in the nature of the spacers has made possible detailed investigation of their binding tendencies towards metal ions and organic and inorganic anions.

Hugo Schiff, his bases, his classroom

Hugo Schiff (Frankfurt am Main, 1834 — Florence, 1915) studied chemistry at the University of Göttingen and prepared his dissertation under the supervision of Friedrich Wöhler. A supporter of socialism and personal friend of Karl Marx, he was inquired by the police and in 1856 had to take refuge in Switzerland, where he continued his studies at the University of Bern. He went back to Göttingen just for discussing his thesis in 1857. He published a short note on what we call today 'Schiff bases' in 1864 (H. Schiff, *Justus Liebigs Ann. Chem.* **1864**, *131*, 118–125, "Mittheilungen aus dem Universitäts-Laboratorium in Pisa" — A report from the University Laboratory in Pisa). In fact, over the period 1863–1864, he was in Pisa as an assistant of Professor Paolo Tassinari. At the end of 1864, he was appointed as a professor at the *Regio Istituto di Studi Superiori Pratici e di Perfezionamento* in Florence, a doctorate school and a centre of advanced studies. From Florence, he published a complete study on 'Schiff bases' and its analytical applications (H. Schiff, *Justus Liebigs Ann. Chem.* **1866**, *140*, 9–55, "Eine neue Reihe organischer Diamine" — A new series of organic diamines), a topic which he did not investigate any longer in his career. Ugo Schiff (living in Italy, he deliberately lost the initial "H" of his first name) was instrumental in the construction of the new Institute of Chemistry in Florence, which replaced the old Grand Duke Stables, close to the Basilica of the Most Holy Annunciation, downtown. He personally designed the main amphitheatre, in which he used to have classes and in which he delivered his 'last lecture' (*Lectio Magistralis*), eternalised by a famous photograph (see a detail in Figure 8.5).

(*Continued*)

(Continued)

Figure 8.5. Ugo Schiff giving his last lecture (*Lectio Magistralis*) in the amphitheatre of the Institute of Chemistry in Florence, on Saturday 24 April 1915. The lecture had been scheduled on Schiff's 81st birthday, Monday 26 April 1915. However, Schiff anticipated the lecture to Saturday, a half-working day, so as to not interfere too much with the regular timetable of classes.

Figure 8.6 displays an inscription in Ancient Greek set over the blackboard, not shown in the detail in Figure 8.5: ΠΑΝΤΑ ΜΕΤΡΩ ΚΑΙ ΑΡΙΘΜΩ ΚΑΙ ΣΤΑΘΜΩ — [You, My God, have ordered] *all things by measure and number and weight.*

Figure 8.6. The Ancient Greek inscription set on the wall over the blackboard in the amphitheatre of the Institute of Chemistry. The sentence (*all things by measure and number and weight*) has been taken from the Book of Wisdom, 11, 20.

(Continued)

(Continued)

The sentence, an invocation by Solomon to God, taken from the Book of Wisdom, Chapter 11, seems to describe the divine order of the physical world and to suggest the scientific keys for studying and interpreting nature. However, on reading the complete paragraph in the Book of Wisdom, the meaning appears totally different and unrelated to science: "Even without these [the Plagues of Egypt], they [the Egyptians] could have been killed at a single breathe, pursued by justice and winnowed by Your mighty spirit. But You, My God, ordered *all things by measure and number and weight*." Thus, Solomon praises the clemence of God in softly punishing the Egyptians, guilty of persecuting the Israelites. Schiff, son of a Jewish family, very probably knew the Book of Wisdom (even if this book is not accepted in the Jewish Bible), mastered Ancient Greek (as well as other five modern and ancient languages) and was aware of the sense of Solomon's invocation. However, he was intrigued by the 'scientific' misinterpretation of the sentence and wanted it to perpetually admonish students attending classes in the amphitheatre (including the writer of these notes).

Significantly, the Latin version of the above sentence (*Omnia in mensura et numero et pondere*) is present in the *Aula Magna* of the Department of Chemistry at the University of Bologna. The Aula was constructed following the will of Giacomo Ciamician (1857–1922). Ciamician was a younger colleague and a friend of Ugo Schiff and was probably inspired by a visit to the Chemical Amphitheatre in Florence.

9

The Formation of Dicopper(II) Bistren Cryptates and the Nature of their Cavity

The most convenient way for studying the formation of dimetallic bistren cryptates in solution is based on potentiometric titration experiments. In a preliminary study, an acidic aqueous solution of the chosen octamine cryptand (e.g. **33**) is titrated with standard NaOH and the protonation constants of the six secondary amine nitrogen atoms (pK_{Ai}) are determined through nonlinear treatment of the titration curve (potential of the glass electrode, mV, vs volume of NaOH, mL). Then, in the second experiment, a solution, e.g. 5×10^{-4} M in **33**, 1×10^{-3} M in $Cu^{II}(ClO_4)_2$ and 10^{-2} M in $HClO_4$, is titrated with standard NaOH. Through nonlinear least-squares processing of the titration curve, the species present at the equilibrium are identified and the corresponding complexation equilibrium constants are determined. On the basis of the values of the protonation and complexation constants, a distribution diagram (% concentration against pH) showing the species present at the equilibrium can be drawn (see Figure 9.1).[74]

Figure 9.2 illustrates the hypothesised structural features of the species that form over the course of the titration.

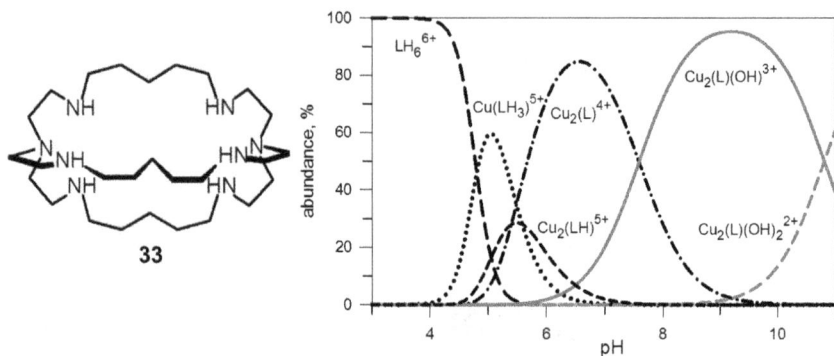

Figure 9.1. Distribution diagram (% concentration with respect to L = 33) of the species at the equilibrium over the 2–12 pH interval for a solution containing 1 equivalent of the bistren cryptand 33 and 2 equivalent of Cu²⁺.[74]

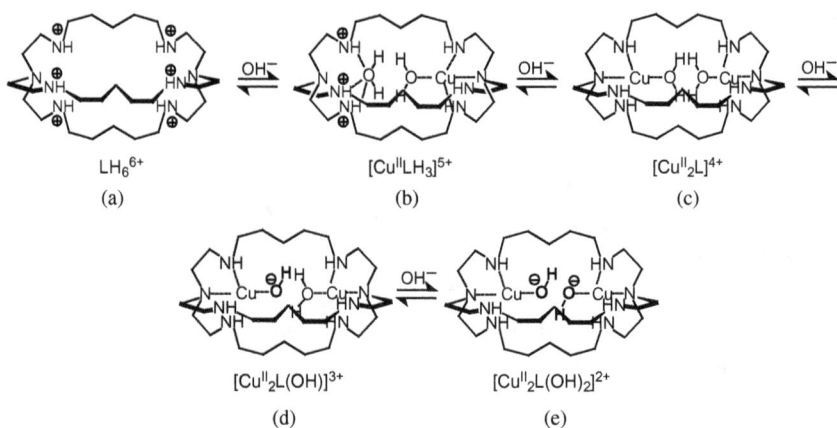

Figure 9.2. Hypothesised structural features of the major species that form during titration of the acidified solution with standard base (L = 33) and described in the distribution diagram in Figure 9.1.

In acidic solutions (pH < 4), the hexaprotonated cryptand LH_6^{6+} (Figure 9.2(a)) is present at 100%. On base addition, LH_6^{6+} disappears, while the mononuclear complex $[Cu^{II}(LH_3)]^{5+}$ forms: in this species, structural formula in Figure 9.2(b), one Cu^{II} ion occupies one tren subunit and completes five-coordination with a water molecule. The three secondary amine nitrogen atoms of the other tren subunit remain protonated and probably establish hydrogen bonding interactions with a second included water molecule, as tentatively sketched in Figure 9.2(b). The $[Cu^{II}(LH_3)]^{5+}$ complex reaches its maximum concentration (60%) at pH = 5. Then, a minor mononuclear species forms, $[Cu^{II}_2(LH)]^{5+}$ (not reported in Figure 9.2), in which the second Cu^{II} ion has entered the second tren subunit, which still keeps one nitrogen atom protonated (30% at pH = 5.5). On moderate pH increase, the last proton is neutralised and the dinuclear complex $[Cu^{II}_2(L)]^{4+}$ forms (Figure 9.2(c)), which contains two equivalent five-coordinate metal centres: it reaches 90% at pH = 6.5. Subsequently, on further pH increase, the two copper(II)-bound water molecules deprotonate stepwise to form, first, the aquo-hydroxo complex $[Cu^{II}_2(L)(OH)]^{3+}$ (95% at pH = 9, Figure 9.2(d)), then the di-hydroxo species $[Cu^{II}_2(L)(OH)_2]^{2+}$ (80% at pH = 12, Figure 9.2(e)).

That displayed by the system copper(II)/**33** is the normal behaviour in solution of a dimetallic bistren cryptate complex, in which the three spacers linking the two tren subunits do not disturb the cavity and its guest, and do not interact with molecules or anions interfacing the two metal centres. Spacers are *inactive*.

However, *active* spacers exist, and one representative example is provided by the classical cryptand **22**. Figure 9.3 shows the distribution of the species at the equilibrium in a solution 5×10^{-4} M in **22** and 1×10^{-3} M in $Cu^{II}(ClO_4)$, over the pH interval 2–12.[74] Also in the present case, in the acidic region, pH < 3, the only species present in solution, with uncomplexed Cu^{2+}, is the hexaprotonated cryptand LH_6^{6+}. On base addition, LH_6^{6+} disappears, while the $[Cu^{II}(LH_3)]^{5+}$ complex forms, whose structure should be the same suggested in the case of cryptand **33**: one tren subunit still triprotonated, the other one hosting a metal ion (Figure 9.2(b)). Then, at pH < 4, both the dinuclear species $[Cu^{II}_2L]^{4+}$ and the mono-hydroxo complex $[Cu^{II}_2L(OH)]^{3+}$ begin to form: $[Cu^{II}_2L]^{4+}$ reaches its maximum concentration, 20%, at pH = 4.5, then disappears, whereas $[Cu^{II}_2L(OH)]^{3+}$ keeps growing to 100% until pH = 6. Noticeably, the mono-hydroxo complex remains the only species in solution until the end of the titration (pH = 11).

The extremely high stability of $[Cu^{II}_2L(OH)]^{3+}$, which minimises the formation of the dinuclear diaquo complex, and prevents the formation of

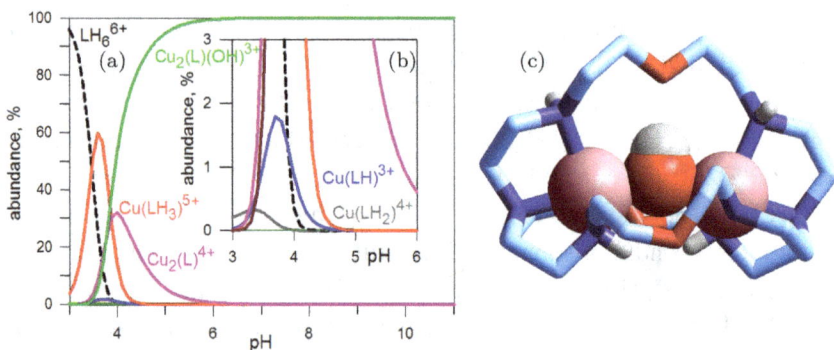

Figure 9.3. (a) Distribution diagram (% concentration with respect to L = **22**) of the species at the equilibrium of a solution 5×10^{-4} M in **22**, 10^{-3} M in $Cu^{II}(ClO_4)_2$ and 10^{-2} M in $HClO_4$ over the course of the titration with standard NaOH[74]; (b) enlarged detail illustrating the formation of the complex species $[Cu^{II}(LH_2)]^{4+}$, grey line, and $[Cu^{II}(LH)]^{3+}$, blue line; (c) the crystal structure of the complex salt $[Cu^{II}_2(\mathbf{22})(OH)]Br_3 \cdot 6H_2O$.[78] All hydrogen atoms have been omitted but that of the included OH^- ion. The hydroxide anion bridges the two Cu^{II} centres and establishes a hydrogen bonding interaction with the facing ethereal oxygen atom of one spacer. The distance $O-H\cdots O$ is 2.01 Å, while the $O\cdots O$ distance in the same fragment is 2.74 Å. The distances of the hydroxide oxygen atom from the other two ethereal oxygen atoms is 4.33 Å.

the di-hydroxo species $[Cu^{II}_2(L)(OH)_2]^{2+}$, can be explained by the crystal structure of the complex salt $[Cu^{II}_2(L)(OH)]Br_3 \cdot 6H_2O$ shown in Figure 9.3(c).[78] In fact, in the $[Cu^{II}_2L(OH)]^{3+}$ complex, an OH^- anion (i) bridges the two Cu^{II} metal ions and (ii) points its hydrogen atom towards a facing ethereal oxygen atom of a spacer. The short $O-H\cdots O$ distance (2.01 Å) is indicative of the existence of a well-defined hydrogen bonding interaction. Moreover, the short $Cu^{II}\cdots Cu^{II}$ distance (3.77 Å) and the value of the $Cu^{II}-O-Cu^{II}$ angle (155°) suggest the occurrence of a pronounced $d_{z2}(Cu1)-2p_z(O)-d_{z2}(Cu2)$ orbital overlap with pairing of the d_{z2} electrons. These two contributions (establishing of the hydrogen bond $O-H\cdots O_{(ethereal)}$ and spin pairing of the two Cu^{II} ions) impart a unique stability to the hydroxo complex. Such a stability is quantitatively expressed by the constant of the anion inclusion equilibrium: $[Cu^{II}_2L]^{4+} + OH^- \rightleftharpoons [Cu^{II}_2L(OH)]^{3+}$, $K = 10^{10.0}$. This value is 5000-fold higher than that observed for the corresponding equilibrium involving cryptand **33**, which contains inactive spacers.[74]

The two examples described above have shown how the spacers of the cryptand affect the acid–base behaviour of the water molecules bound to the Cu^{II} ions of the corresponding dinuclear cryptate. A further distinctive example is provided by the dicopper(II) cryptate of **28**. Figure 9.4(a) displays the distribution diagram of the species at equilibrium in a solution 5×10^{-4} M in **28** and 10^{-3} M in Cu^{II}(ClO$_4$), over the pH interval 2–12.[79]

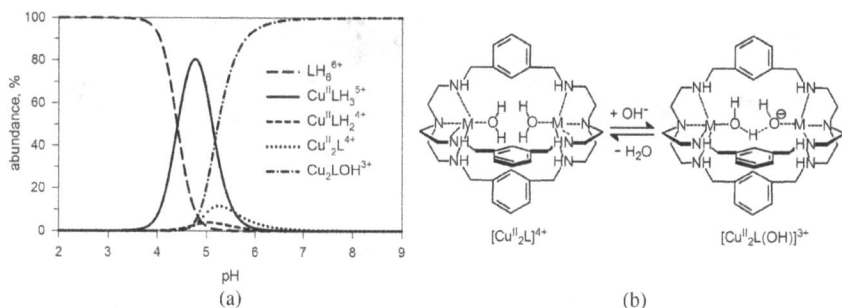

Figure 9.4. (a) Distribution diagram (% concentration with respect to L = **28**) of the species at equilibrium over the 2–9 pH interval, for a solution containing 1 equivalent of the bistren cryptand **28** and 2 equivalent of Cu^{2+};[79] (b) hypothesised coordinative arrangement of the species [Cu$^{II}_2$(L)(H$_2$O)$_2$]$^{4+}$(diaquo) and [Cu$^{II}_2$(L)(OH)]$^{3+}$ (aquo-hydroxo).

Also in the present case, at pH < 4, the hexaprotonated cryptand LH$_6^{6+}$ is present at 100%. On addition of NaOH, the three secondary ammonium groups of one tren subunit are protonated, while the other uptakes a Cu^{II} ion to give the mononuclear complex [Cu^{II}(LH$_3$)]$^{5+}$, which reaches its maximum concentration (80%) at pH = 4.8 and for which the structure of Figure 9.2(b) is suggested. On further base addition, the second tren subunit is also deprotonated and binds a second metal ion, to give the dinuclear species, in which the two Cu^{II} ions show the same coordination sphere: [Cu$^{II}_2$(L)(H$_2$O)$_2$]$^{4+}$. However, this species is formed in a low concentration (max 10% at pH = 5.2), due to the pronounced tendency of one of the two coordinated water molecules to deprotonate, to give the hydroxo complex [Cu$^{II}_2$(L)(OH)]$^{3+}$, present at 100% over pH > 6. This behaviour seems similar to that observed for system **22**. Quite disappointingly, structural data on the [Cu$^{II}_2$(**28**)]$^{4+}$ and [Cu$^{II}_2$(**28**)(OH)]$^{3+}$ complexes are not available. However, there exist crystallographic data on the corresponding complexes of the bistren cryptand **29**, which contains 1,4-xylyl instead of 1,3-xylyl spacers. It is assumed that this small structural difference should not significantly alter coordination and properties of corresponding dicopper(II) cryptates. The structures of the dinuclear complexes of **29** are shown in Figure 9.5.

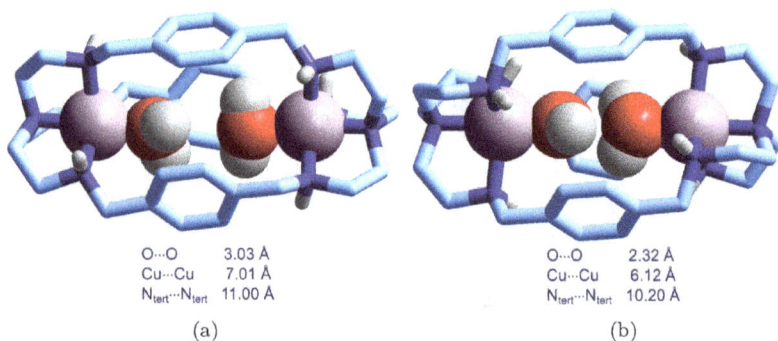

O···O	3.03 Å		O···O	2.32 Å
Cu···Cu	7.01 Å		Cu···Cu	6.12 Å
$N_{tert}···N_{tert}$	11.00 Å		$N_{tert}···N_{tert}$	10.20 Å

(a)   (b)

Figure 9.5. The crystal structures of (a) $[Cu^{II}_2(L)(H_2O)_2](CF_3SO_3)_4 \cdot MeCN$,[80] and (b) $[Cu^{II}_2(L)(H_2O)(OH)](ClO_4)_3$ (L = **29**)[81]; C–H hydrogens, and solvating molecules have been omitted for clarity.

First, it has to be observed that in the $[Cu^{II}_2(L)]^{4+}$ species, two water molecules are included in the cavity, and each molecule completes the five-coordination of each metal centre (see Figure 9.5(a)).[80] Such a structural arrangement had also been hypothesised for corresponding complexes of **33** and **22** (Figure 9.2(c)). The $[Cu^{II}(L)OH]^{3+}$ cryptate (L = **29**, Figure 9.5(b)) contains a metal-bound water molecule and a metal-bound hydroxide ion.[81] However, while in the diaquo complex the two water molecules are well separated, in the aquo-hydroxo species water and hydroxide strongly interact, which induces a significant rearrangement of the cryptate framework. In particular, in the diaquo complex (Figure 9.5(a)), the distance of the oxygen atoms of the two water molecules is 3.03 Å, whereas in the aquo-hydroxo complex (Figure 9.5(b)) the distance between the aquo oxygen atom and the hydroxo oxygen atom is 2.32 Å. To such a decrease, drastic reductions of the $Cu^{II}---Cu^{II}$ distance (from 7.01 to 6.12 Å) and of the $N_{tert}---N_{tert}$ distance (from 11.00 to 10.20 Å) correspond. The short O---O distance indicates that a strong hydrogen bond is established between one hydrogen atom of the coordinated water molecule, whose partial positive charge has been increased by metal coordination, and the hydroxo oxygen atom, which detains a formal negative charge. The $H_2O\cdots OH^-$ species can be considered an encapsulated form of monohydrated hydroxide ion, $H_3O_2^-$. The structure of $H_3O_2^-$ in the gas phase, calculated through DFT studies,[82] is sketched in Figure 9.6(a).

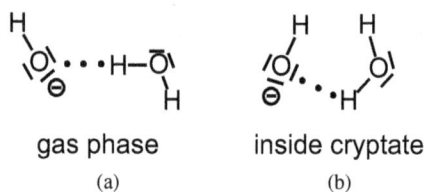

gas phase   inside cryptate

(a)   (b)

Figure 9.6. Structural arrangement of the monohydrated hydroxide ion $H_3O_2^-$; (a) as calculated in the gas phase,[82] and (b) as observed inside the dicopper(II) cryptate.[81]

In the gaseous $H_3O_2^-$, the O---O distance is 2.54 Å,[82] a value distinctly higher than that observed in the $H_3O_2^-$ species bound to two Cu^{II} ions and included in the cryptate (2.32 Å, structural formula sketched in Figure 9.6(b)). This suggests that the encapsulation of the $H_3O_2^-$ complex in the dimetallic cryptate results in a significant increase of stability, which may be related to the enhanced H-bond donicity of the water –OH fragment, when bound to a metal centre. The energy gain associated with the hydrogen bond formation more than compensates for the conformational energy loss associated with the contraction of the cryptate along its longitudinal axis. Noticeably, the monohydrated form of OH^- is not stable enough to be observed and characterised in aqueous solution or in the solid state. The dicopper(II) cryptate provides to $H_3O_2^-$ a safe shelter, where to live a long and comfortable life.

At this stage, it can be surprising that in the $[Cu^{II}_2(L)(H_2O)_2]^{4+}$ complex (L = **28, 29**), the deprotonation of one water molecule induces a drastic conformational rearrangement and simultaneous formation of a bridging $H_3O_2^-$ anion, whereas in the $[Cu^{II}_2(L)(H_2O)_2]^{4+}$ (L = **33**) nothing similar takes place, but metal-bound water molecules deprotonate independently and without any reciprocal interference. In fact, the spacers of the considered cryptands have a comparable length and, therefore, their cavities should have similar shape and volume. A possible explanation is probably related to the nature of the spacers: aliphatic in cryptand **33** and aromatic in cryptands **28** and **29**. It is suggested that the OH⁻ oxygen atom of the encapsulated $H_3O_2^-$ anion establishes π-donor-π-acceptor interactions with the phenyl ring of the spacers. Indeed, in the $[Cu^{II}_2(L)(H_2O)(OH)]^{3+}$ complex taken as a model (L = **29**), each one of the three phenyl rings is oriented in a parallel mode with respect to the cavity in which $H_3O_2^-$ is included, thus being prone to receive electron density through a π mechanism: $p_z(O) \rightarrow \pi^*(phenyl)$.

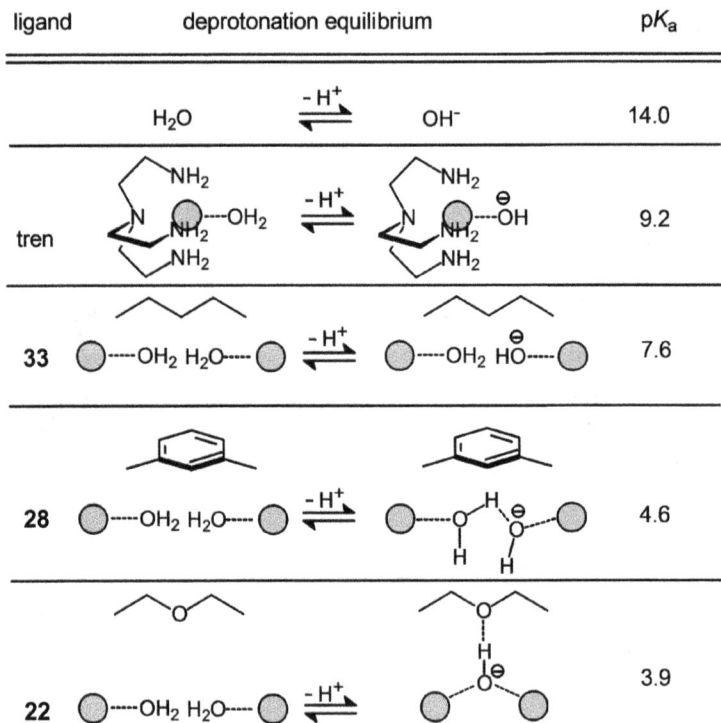

Figure 9.7. Deprotonation equilibria involving H_2O in bulk water, bound to $[Cu^{II}(tren)]^{2+}$ and included in dicopper(II) complexes of bistren cryptands **22**, **28** and **33**, with the corresponding pK_A values. Grey spheres symbolise Cu^{II} ions; structural formulae of the spacers of the cryptands are sketched; pK_A of $[Cu^{II}(tren)(H_2O)]^{2+}$ from Ref. 83.

The deprotonation equilibria involving water molecules included in dicopper(II) cryptates are summarised and pictorially illustrated in Figure 9.7, and corresponding pK_A values are also displayed. Bulk water and the $[Cu(tren)H_2O]^{2+}$ complex are also considered for comparative purposes.

Axial coordination to a metal-tetramine subunit — $[Cu^{II}(tren)]^{2+}$ — highly enhances the acidity of the water molecule. In fact, electron density is transferred from coordinated H_2O to the metal, which makes the partially positive charge on hydrogen atoms increase, thus favouring their release as protons. This is quantitatively expressed by the decrease of pK_A from 14 (pure water) to 9.2 $[Cu^{II}(tren)(H_2O)]^{2+}$. Then, on moving to bistren dinuclear cryptates, it may seem surprising that the water molecule coordinated to a Cu^{II}-tren subunit of cryptand **33** is distinctly more acidic ($pK_A = 7.6$) than observed in the monomeric counterpart $[Cu^{II}(tren)]^{2+}$. However, this can be explained by considering two distinct effects; (i) water deprotonation in the $[Cu^{II}_2(33)(H_2O)_2]^{4+}$ complex is favoured by a statistical factor of 2 (the H^+ ion can be released by *two* water molecules), which gives the moderate advantage of 0.3 log units in pK_A; and (ii) most importantly, the two water molecules inside the cryptate are facing each other and suffer from electrostatic (and steric) repulsions between the partially positive hydrogen atoms. On deprotonation, electrostatic repulsions are cancelled, which makes acidity increase. On the other hand, the acidity of water molecules included in $[Cu^{II}_2(L)(H_2O)_2]^{4+}$ (L = **28**, $pK_A = 4.6$) and $[Cu^{II}_2(L)(H_2O)_2]^{4+}$ (L = **22**, $pK_A = 3.9$) is significantly enhanced by the favourable contribution resulting from the interaction of the conjugate base ($H_3O_2^-$ and OH^-, respectively) with the spacer(s). $[Cu^{II}_2(L)(H_2O)_2]^{4+}$ (L = **22**) in particular appears as a Brønsted acid of respectable strength, higher, for instance, of that of acetic acid — a further manifestation of the cryptate effect.

10

Anion Recognition by Dicopper(II) Bistren Cryptates: The Geometrical Factor

Studies illustrated in Chapter 9 have demonstrated that dimetallic bistren cryptates do not have a 'void' cavity ready to include anions. The cavity is always filled with water molecules or their conjugate bases, OH⁻ or $H_3O_2^-$. Thus, in an aqueous solution, anion inclusion is in any case preceded by the removal of H_2O/OH^-, a process that costs significant energy. Moreover, since studies are typically carried out in a neutral solution (pH around 7), anion inclusion involves the replacement by the anion X⁻ of OH⁻, hydrated or not. In any case, whatever the included ligand is, equilibrium studies at a given pH are expected to disclose selectivity of anion inclusion depending upon size–shape matching of receptor and substrate. A convincing example is provided by the dicopper(II) cryptate of bistren **28**.

To verify the affinity of the dicopper(II) cryptate for a given anion X⁻, an aqueous solution 5×10^{-4} M in **28** and 10^{-3} M in Cu(ClO₄)₂, buffered at pH = 8, was titrated with a standard solution of NaX. At pH 8, the equilibrium under investigation involves the replacement of the included $H_3O_2^-$ ($H_2O\cdots OH$)$^-$ by the ambidentate anion X^-, as described by eq(1):

$$[Cu^{II}_2(L)(H_2O)(OH)]^{3+} + X^- \leftrightharpoons [Cu^{II}_2(L)(X)]^{3+} + OH^- + H_2O \quad (10.1)$$

Let us consider first the titration with the azide anion. On addition of N_3^- to a buffered solution of the cryptate, the colour turns from pale blue to bright green. The intense colour developed results from the formation of a band at 400 nm. Figure 10.1(a) shows the family of spectra taken over the course of the titration with azide.

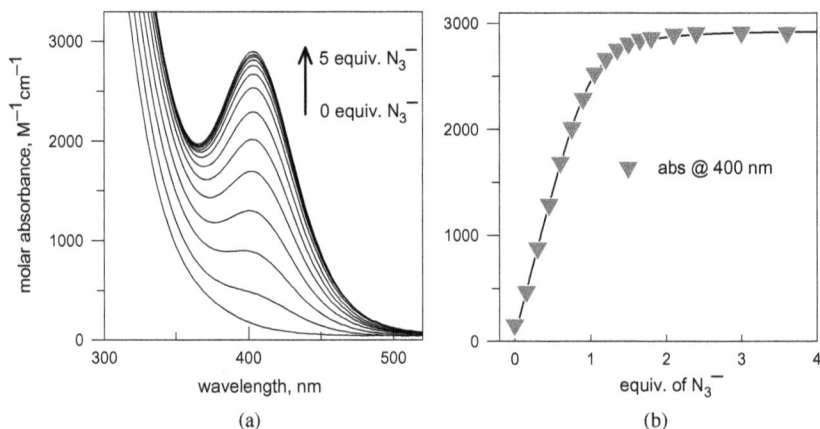

Figure 10.1. (a) Spectra taken over the course of the titration of a solution of the dicopper(II) complex of cryptand **28**, buffered at pH 8, with a standard solution of NaN₃; the intense band which develops results from a charge transfer transition from the encapsulated azide to the copper(II) ion; (b) titration profile (absorbance at 400 nm against equivalent of N_3^-), indicating the 1:1 stoichiometry of the inclusion process; from nonlinear least-squares treatment of titration data, a log K value of 4.78 ± 0.05 was calculated for the inclusion equilibrium (10.1).

Figure 10.1(b) shows the titration profile (absorbance of the band centred at $\lambda_{max} = 400$ nm against equivalent of N_3^-). The profile clearly indicates the 1:1 stoichiometry of the receptor/anion complex that forms. From nonlinear least-squares treatment of the spectrophotometric data over the 350–500 nm interval, a log K value was calculated (4.78 ± 0.05)

for the equilibrium (10.1). The 1:1 stoichiometry suggests the formation of an inclusion complex. Such a hypothesis was substantiated by X-ray diffraction studies on a single crystal. Figure 10.2(a) shows the structure of the $[Cu^{II}_2(L)(N_3)]^{3+}$ complex (L = **28**).

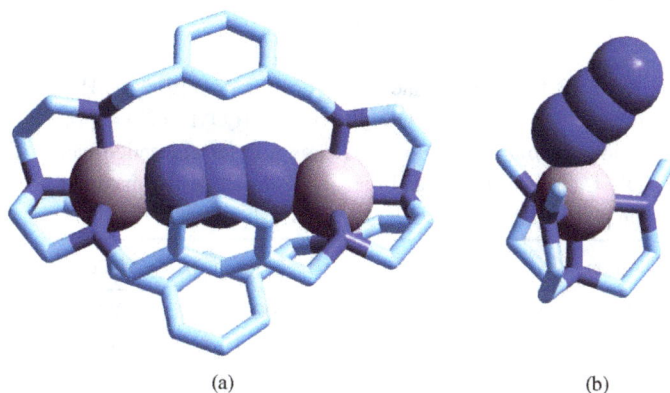

(a)   (b)

Figure 10.2. (a) Crystal structure of the ternary complex salt $[Cu^{II}_2(L)(N_3)](ClO_4)_3 \cdot$ EtOH·MeCN, L = **28**; (b) crystal structure of the $[Cu^{II}(Me_3tren)(N_3)]ClO_4$ (Me_3tren = tris(2-(N-methylamino)ethyl)amine). Hydrogen atoms, perchlorate ions and solvating molecules have been omitted for clarity.

It is observed in Figure 10.2(a) that the ambidentate N_3^- anion is encapsulated by the cryptand and bridges the two Cu^{II} centres. Noticeably, the azide ion is bound to each copper(II) ion in an 'unnatural' linear mode. In fact, the 'natural' mode is 'bent', as observed for instance in all the ternary complexes of azide with a copper(II) tetramine moiety, in the absence of steric constraints. As an example, Figure 10.2(b) shows the structure of the $[Cu^{II}(Me_3tren)(N_3)]^+$ complex (Me_3tren = tris(2-(N-methylamino)ethyl) amine), in which the Cu–N–N angle is 128°. This value reflects the predominant sp^2 character of the terminal nitrogen atom of azide. The collinear arrangement of N_3^- in the cryptate complex is sterically imposed by the ligand and may have a cost in terms of energy of the coordinative bond. It will be shown later in this chapter that such an energy cost does not prevent the formation of an especially stable complex (compared to other anions).

Similar titration experiments were carried out for a variety of polyatomic anions. In all cases, anion addition caused a sharp spectral change (in general, appearance and development of the anion-to-metal charge transfer band in the UV region), and absorbance against anion equivalent plots indicated the formation of a complex of 1:1 stoichiometry. Log K values of formation equilibria at pH 8 and some spectral features of the ternary complexes $[Cu^{II}_2(L)(X)]^{3+}$ (L = **28**) are reported in Table 10.1.

Table 10.1. Log K values for the equilibrium $[Cu^{II}_2(L)(H_2O)(OH)]^{3+} + X^- \leftrightarrows [Cu^{II}_2(\mathbf{31})(X)]^{3+} + OH^- + H_2O$ (L = **28**) at pH = 8 and spectral features of the charge transfer absorption band (X^--Cu^{II}), wavelength and molar absorbance at the maximum (in parentheses uncertainties on log K values).

Anion, X^-	log K	In λ_{max}, nm	ε, M^{-1} cm^{-1}
N_3^-	4.78 (0.05)	400	3895
NCO^-	4.60 (0.05)	340	4904
NCS^-	2.95 (0.05)	352	2696
SO_4^{2-}	3.26 (0.08)	348	2585
$HCOO^-$	3.32 (0.04)	330	4523
CH_3COO^-	2.97 (0.05)	348	3166
HCO_3^-	4.56 (0.04)	360	3733
NO_3^-	2.70 (0.06)	348	3483

The log K values demonstrate that the dimetallic cryptate, with its ellipsoidal cavity, is a very appropriate receptor for the rod-like N_3^- and NCO^- anions, for which the highest log K values are observed. Indeed, X-ray diffraction studies showed that the cyanate-containing cryptate shows a coordinative arrangement similar to that observed for the corresponding azide complex, with the anion encompassing the two metal centres, according to a collinear mode (see Figure 10.3(a)).

However, the other rod-like anion NCS^- forms a complex much less stable than the other pseudohalides N_3^- and NCO^-. Such a behaviour may be surprising from the classical coordination chemistry point of view, considering that NCS^- is higher in the spectrochemical series than NCO^- and N_3^-. On the other hand, it is worth noting that the Y-shaped HCO_3^- ion, which typically displays poor coordinating tendencies towards transition

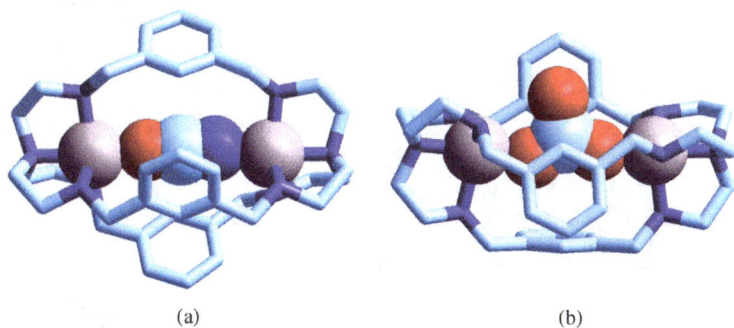

(a)   (b)

Figure 10.3. Crystal structures of (a) the $[Cu^{II}_2(L)(NCO)](ClO_4)_3 \cdot EtOH \cdot MeCN$ complex salt, and of (b) the $[Cu^{II}_2(L)(HCO_3)](ClO_4)_3 \cdot MeCN$ [a complex salt (L = **28**). Hydrogen atoms, perchlorate counterions and solvating molecules have been omitted for clarity.

metals, forms a very stable inclusion complex. Such a high affinity is also demonstrated by the fact that a neutral or slightly basic solution of the cryptate, when exposed to air, turns green in colour in a few minutes due to carbon dioxide uptake and HCO_3^- incorporation. The X-ray structure of the crystalline complex salt $[Cu^{II}_2(L)(HCO_3)](ClO_4)_3 \cdot MeCN$ (L = **28**, Figure 10.3(b)) shows that two oxygen atoms of HCO_3^- bridge the two Cu^{II} ions, lying on the line that joins the two metal centres.

The above findings suggest that the stability of the inclusion complex is neither related to the shape of the anion nor to its intrinsic donating tendencies. Rather, it seems connected to the anion's capability to place two of its donor atoms in the fifth coordination site of each Cu^{II} centre, thus encompassing the intermetallic distance. In this regard, it is useful to consider the 'bite' of each anion, i.e. the distance between two proximate donor atoms. In Figure 10.4(a) the bite lengths of NCO^-, HCO_3^- and SO_4^{2-} are considered as examples.

Then, log K values were plotted against the bite length (see Figure 10.4(b)), disclosing the existence of a sharp *peak selectivity*. The linear N_3^- and NCO^- anions and the Y-shaped HCO_3^- anion show the highest affinity because their bite fits well the distance between the two Cu^{II} 'fifth coordination sites', without disturbing the cryptate framework, relaxed to its lowest energy conformational arrangement. On the other hand, the linear NCS^- ion has too long a bite and its

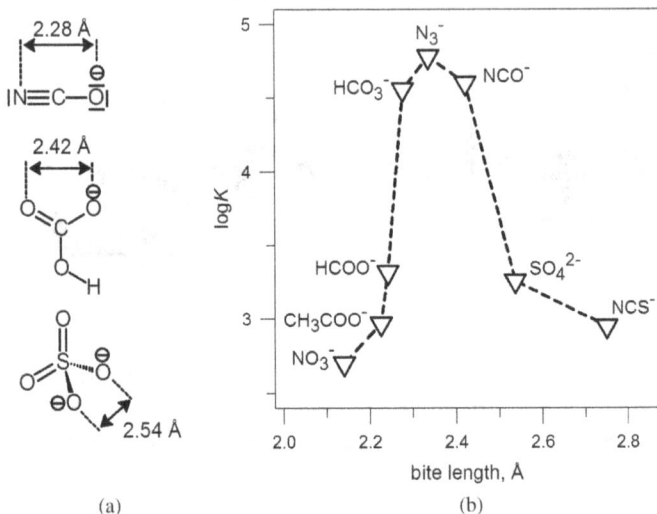

Figure 10.4. (a) Definition of bite: the distance between two consecutive donor atoms in an ambidentate anion (cyanate, hydrogen carbonate, sulphate); (b) plot of log K values (from Table 10.1) against the bite length of some ambidentate anions.

inclusion should induce an endergonic rearrangement of the cryptate and of its organic backbone. In a similar way, the triangular NO_3^- has too short a bite and forces the cavity to contract, which induces a conformational rearrangement again and makes log K decrease. Thus, it is a purely geometrical factor that dominates complexation and overcomes all the effects related to the anion's shape, coordinative tendencies and charge. In particular, the sulphate anion should be favoured in principle by the double negative charge, but it is bound weakly to the dimetallic cryptate because it presents too large a bite.

Finally, it must be emphasised that log K values reported in Table 10.1 and in Figure 10.4(b) do not refer to thermodynamic constants, but to 'conditional' constants, i.e. to quantities strictly related to the environmental conditions of the experiment, in this case to pH. The choice of pH of the solution is critical. In fact, the experimentally determined constant K is related to the thermodynamic constant of the anion exchange equilibrium (10.1), K^* through the relationship $K = K^*/[OH^-]$. Thus, a pH decrease would make K values increase: for instance, at pH = 7, where

$[Cu^{II}_2(L)(H_2O)(OH)]^{3+}$ is still the dominating species (>98%), log K values are about one unit larger than at pH = 8. However, at pH = 7, anions presenting the highest affinity for the cryptate (e.g. N_3^- and NCO^-) exhibit log K values close to the upper limit (or beyond) for a safe spectrophotometric determination. In any case, the selectivity pattern displayed in Figure 10.4(b) is maintained.

It has to be noted that $[Cu^{II}_2(L)(H_2O)(OH)]^{3+}$ (L = **28**) *does not show any affinity towards halide ions*: in particular, even large additions of sodium halides do not modify the absorption spectrum of the dicopper(II) cryptate. Halides are ambidentate anions and, under the appropriate conditions, can bridge two coordinatively unsaturated Cu^{II} ions. However, in the present case, bridging would require a drastic contraction of the cavity and a severe rearrangement of the cryptate framework. Such an endergonic process is not compensated by the formation of the Cu^{II}–X–Cu^{II} bonds. It derives that the dicopper(II) cryptate $[Cu^{II}_2(L)(H_2O)(OH)]^{3+}$ is not tempted to replace with a halide the $H_3O_2^-$ ion, well settled in the cavity.

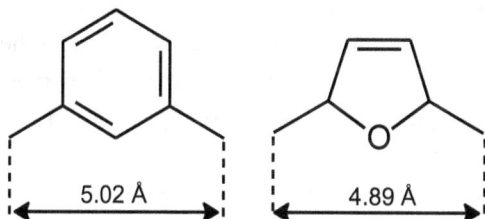

Figure 10.5. C–C distance in the spacers of cryptands **28** and **31**, as observed in metal-free cryptands.

It is therefore surprising that the dicopper(II) complex of cryptand **31**, whose spacer, a 2,5-dimethylfuran fragment, is only 0.13 Å shorter than that of the 1,3-xylyl spacer of **28** (see Figure 10.5), is able to incorporate halide ions, forming stable complexes in water. X-ray diffraction studies on the crystalline salts $[Cu^{II}_2(L)(Cl)](ClO_4)_3$ and $[Cu^{II}_2(L)(Br)](ClO_4)_3$ (L = **31**) demonstrated that chloride and bromide ions are really encapsulated by the cryptate and bridge the two Cu^{II} centres, as shown in Figure 10.6.

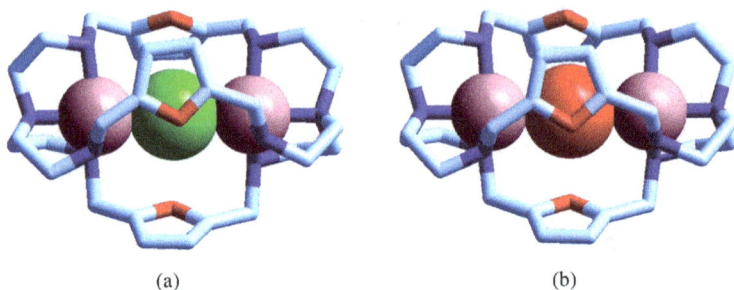

(a)   (b)

Figure 10.6. The crystal and molecular structure of: (a) $[Cu^{II}_2(L)(Cl)](ClO_4)_3$ and (b) $[Cu^{II}_2(L)(Br)](ClO_4)_3$, L = **31**. Hydrogen atoms and perchlorate counterions have been omitted for clarity.

The distances between tertiary amine nitrogen atoms in the two inclusion complexes (8.74 Å for Cl⁻ and 8.96 Å for Br⁻, L = **31**) are noticeably smaller than observed in the $[Cu^{II}_2(L)(N_3)]^{3+}$ cryptate (10.09 Å, L = **28**), which suggests the occurrence of a marked contraction of the bistren framework. The corresponding conformational energy cost seems to be compensated, other than by the formation of the Cu–X–Cu bonds, by an additional term associated with the interaction between the halide and the oxygen atoms of the furan rings. In fact, the X···O distances (3.17 ± 0.03 Å for Cl⁻ and 3.19 ± 0.02 Å for Br⁻) are appreciably lower than the sum of the van der Waals radii (3.27 and 3.37 Å, respectively). Such interactions have a profound influence on energy and intensity of the absorption spectra of the halide inclusion complexes, as will be discussed below.

Potentiometric titration experiments on the $Cu^{2+}/$**31** system in a molar ratio 2:1 lead to the distribution diagram illustrated in Figure 10.7(a). Two major dimetallic species are present over the 3–11 pH interval: (i) the $[Cu^{II}_2(L)]^{4+}$ complex predominant in the 4–7 pH range (maximum concentration: 85% at pH = 5, red line in Figure 10.7(a), pale blue colour of the solution), and (ii) the $[Cu^{II}_2(L)(OH)]^{3+}$ complex, present in the 5–11 pH range (maximum concentration: 97% at pH = 8.5, blue line, intense emerald green colour and an absorption band centred at 362 nm; molar absorbance, $\varepsilon = 6500$ M⁻¹ cm⁻¹). The encapsulation of the small hydroxide ion requires a severe contraction of the bistren framework. This is demonstrated by the crystal and molecular structure of the complex salt

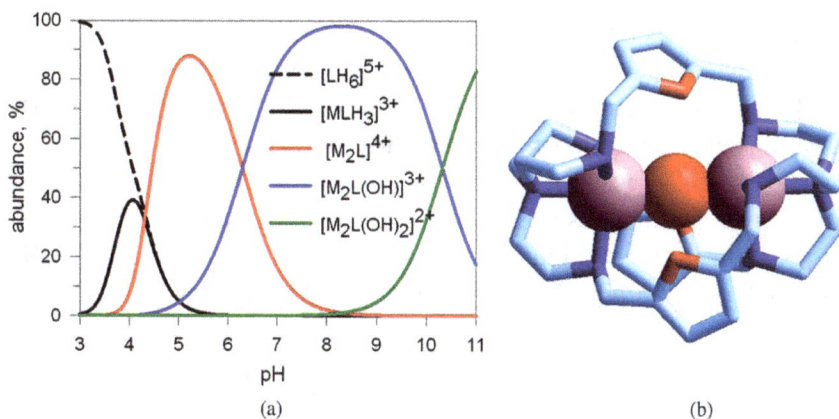

(a)   (b)

Figure 10.7. (a) Distribution diagram (% concentration with respect to L = **31**) of the species present at the equilibrium for the system **31**/Cu^{II} (ligand-to-metal ratio 1:2), in an aqueous solution 0.1 M in $NaClO_4$, at 25°C; (b) the crystal structure of the complex salt $[Cu^{II}_2(L)(OH)](CF_3SO_3)_3$ (L = **31**). All hydrogen atoms have been omitted, including that of the hydroxide ion. The oxygen atom of OH^- establishes a well-defined interaction with the oxygen atoms of the three furan rings of the cryptand: in particular, the O(hydroxide)$\cdots$O(furan) distances are 2.87 ± 0.05 Å, distinctly lower than twice the van der Waals radius of oxygen (3.04 Å).

$[Cu^{II}_2(L)(OH)](CF_3SO_3)_3$ shown in Figure 10.7(b). In fact, the N_{tert}---N_{tert} distance (8.05 Å) is especially low, as also observed for the dicopper(II) complex of cryptand **31**, when chloride and bromide ions are encapsulated. In the hydroxo-cryptate a well-defined interaction between the oxygen atom of the included hydroxide and the ethereal oxygen atoms of the furan rings exists, which is demonstrated by the fact that the measured O$\cdots$O distance is appreciably smaller than the sum of the van der Waals radii (2.87 vs 3.04 Å). It is this additional interaction that probably compensates the conformational energy cost associated with the cryptand's rearrangement. Note that in the analogous complex $[Cu^{II}_2(L)(OH)]^{3+}$ (L = **22**), whose structure has been shown in Figure 9.3(c) and in which a hydrogen bonding interaction between the OH^- ion and one of the ethereal oxygen atoms of the ligating framework is operating, the N_{tert}---N_{tert} distance (7.68 Å) is even shorter than in the $[Cu^{II}_2(L)(OH)]^{3+}$ hydroxo-cryptate (L = **31**). Finally, it has to be recalled that in both $[Cu^{II}_2(L)(OH)]^3$ cryptates of **22** and **31**, the interaction of the two Cu^{II} centres mediated by the hydroxo bridge and consequent spin pairing further contribute to stability.

In order to determine the affinity of the [Cu^{II}$_2$(L)]$^{4+}$ cryptate (L = **31**) towards halide ions, a solution containing the bistren cryptand and two equivalent of Cu^{II} was adjusted to pH = 5.25 with MES buffer and was titrated with a standard solution of a sodium halide, NaX. On X$^-$ addition (X = Cl, Br, I), the pale blue solution (containing as major species, 87%, the 'void' complex [Cu^{II}$_2$L]$^{4+}$) took an intense bright yellow colour.

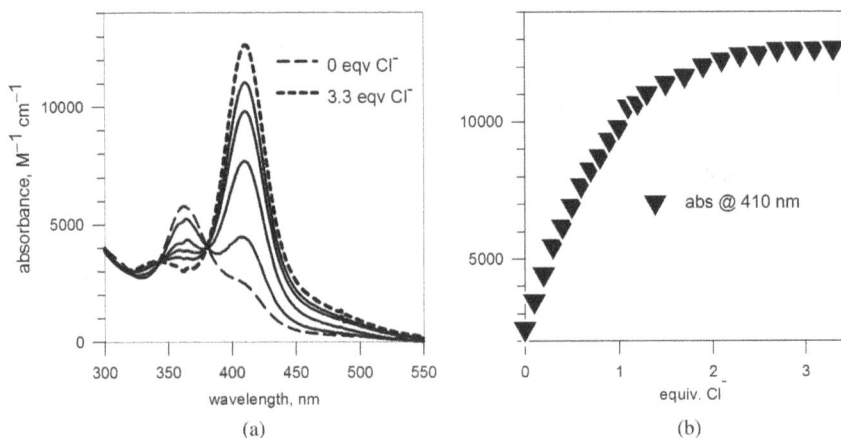

Figure 10.8. (a) Selected spectra recorded over the course of the titration of a solution containing bistren cryptand **31** + 2 equivalent of Cu^{2+} with a standard solution of NaCl. The solution had been buffered to pH 5.25, where the [Cu^{II}$_2$(L)]$^{4+}$ species is present at 87%. The increasing band centred at 410 nm corresponds to the formation of the [Cu^{II}$_2$(L)(Cl)]$^{3+}$ inclusion complex; (b) titration profile at 410 nm. Nonlinear least-squares treatment of spectrophotometric data gave a log $K = 3.98 \pm 0.02$ for the equilibrium: [Cu^{II}$_2$(L)]$^{4+}$ + Cl$^-$ ⇆ [Cu^{II}$_2$(L)(Cl)]$^{3+}$.

Figure 10.8(a) reports some selected spectra taken over the course of the titration with the chloride ion. The titration profile in Figure 10.8(b) (molar absorbance at $\lambda_{max} = 410$ nm against equivalent of Cl$^-$) corresponds to the formation of the 1:1 inclusion complex, whose structure has been shown in Figure 10.6(a). The log K value (conditional constant at pH = 5.25) for the equilibrium (10.2):

$$[Cu^{II}_2(L)]^{4+} + Cl^- = [Cu^{II}_2(L)(Cl)]^{3+} \qquad (10.2)$$

as obtained through a nonlinear least-squares processing of titration data is 3.98 ± 0.02. The intense band at 410 nm (ε: 12600 M^{-1} cm^{-1}) has a ligand-to-metal charge transfer (LMCT) nature. The low energy of the band and its high absorbance (responsible for the intense bright yellow colour) are unusual for a charge transfer from Cl$^-$ to Cu^{II} in a tetramine complex. The interaction of chloride with the cryptand's ethereal oxygen atoms may raise the energy of the filled Cl$^-$ orbital from which the electron is excited to the half-filled d orbital of Cu^{II}, which accounts for the relatively low energy of the transition, responsible for the bright yellow colour. The unusually large value of the absorbance of $[Cu^{II}_2(L)(Cl)]^{3+}$ results from the high probability of the charge transfer process, from *three* MO orbitals (of chloride and oxygen mixed nature) to *two* metal orbitals, six times that observable in the $[Cu^{II}(tren)Cl]^{2+}$ complex (occurring at a much lower wavelength).

Similar titration profiles were obtained for bromide and iodide and the corresponding conditional constants (at pH = 5.25) were calculated: Br$^-$, log $K = 3.01 \pm 0.01$, $\varepsilon = 10800$ M^{-1} cm^{-1}; I$^-$, 2.39 ± 0.02, 950 cm^{-1} M^{-1}. Surprisingly, titration with NaF did not cause any development of the typically observed yellow colour. However, it must be considered that the energy of the LMCT transition increases with the increasing electronegativity of the halogen atom. Thus, the fluoride-to-copper(II) transition, due to the high electronegativity of fluorine, is probably shifted to lower wavelengths, in particular in the UV portion of the spectrum, where it may be obscured by the strong amine-to-metal CT transitions. However, significant spectral modifications in the UV region observed during the titration with NaF of the cryptate solution, buffered at pH = 5.25, generated a saturation profile and the pertinent log K value could be determined: 3.20 ± 0.02.

The plot in Figure 10.9(a) (log K against halide ionic radius) shows a defined selectivity in favour of chloride: log K values (at pH = 5.25) range within an interval of 1.2 log units. In Figure 10.9(b) the log K against ionic radius plot is compared with the log K against bite length plot for polyatomic anions already illustrated in Figure 10.4(b) (L = **28**). Values of K for the two systems cannot be compared because they refer to solutions buffered at different pH values. However, bite length selectivity is higher (the plot is narrower) than halide radius selectivity. This may be due to the fact that bite selectivity has a purely geometric nature,

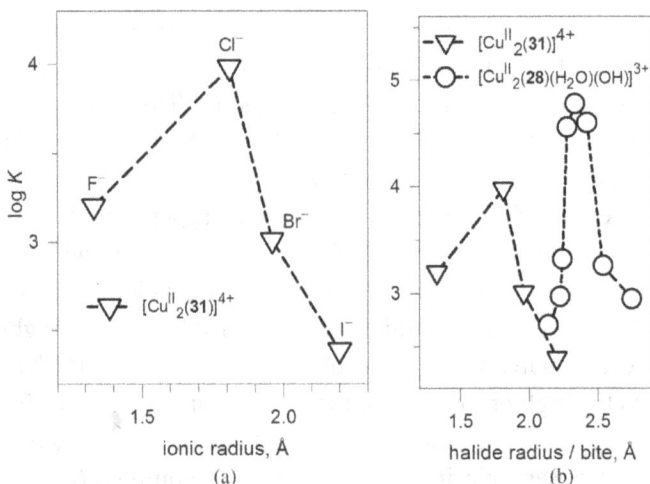

Figure 10.9. (a) Selectivity pattern for the affinity of halides towards the dimetallic cryptate receptor $[Cu^{II}_2(L)]^{4+}$ (L = **31**); log K values refer to the equilibrium: $[Cu^{II}_2(L)]^{4+}$ + $X^- \leftrightarrows [Cu^{II}_2(L)(X)]^{3+}$ (L = **31**) at pH = 5.25 (X^- = halide anion); (b) comparison with the log K against bite length plot for polyatomic anions of Figure 10.4(b) (L = **28**, conditional constants at pH = 8).

whereas halide selectivity results from the several contributions including the furane–halide interaction, permeability of the halide to the Cu^{II}–Cu^{II} communication, spin-pairing extent and matching of the sizes of the halide and receptor cavity.

In any case, the $[Cu^{II}_2(L)]^{4+}$ cryptate (L = **31**) is special, as it has capability to include anions of varying size and shape, mono- and polyatomic. In fact, it encapsulates not only the small halides and hydroxide ions, but also linear triatomic anions like N_3^- and NCS^-. In particular, when titrating with NaN_3 a solution of $[Cu^{II}_2(L)]^{4+}$, buffered to pH = 5.25, the pale blue solution changes to an intense olive green colour: the plot of the absorbance of the band which developed at 386 nm against equivalent of N_3^- indicated the formation of a 1:1 adduct with a log $K = 4.70 \pm 0.06$, for an equilibrium of type (10.2). The development of a CT band (ε = 6100 M^{-1} cm^{-1}) as well as the 1:1 stoichiometry strongly suggest that the N_3^- ion is included within the cage. A similar behaviour was observed with NCS^- (log K = 4.28 ± 0.03). Thus, it appears that the $[Cu^{II}_2(L)]^{4+}$ complex (L = **31**) displays a pronounced versatility, being able to contract/expand its cavity at will in order to include anions of variable size and shape, from the small hydroxide and fluoride to the large thiocyanate. This does not depend on an intrinsic flexibility of the 2,5-dimethylfuran spacers (the aromatic furan ring exhibits the same rigidity of the phenyl ring), but on the establishing of bonding interactions between the furan oxygen atoms and the included ion, either halide or hydroxide, whose energy compensates that spent for the contraction of the cryptand framework.

11

Anion Fluorescence Sensing by Dimetallic Bistren Cryptates: The Fluorophore–Spacer–Receptor Paradigm

Recognition refers to the successful selective interaction of the receptor (e.g. a 'void' dimetallic cryptate) with a substrate (e.g. an anion), in the presence of other competing substrates. Recognition can be monitored by the operator through a variety of instrumental responses, e.g. the shift of an NMR signal or of the electrochemical potential, a modification of the UV–vis absorption spectrum (accompanied by a colour change), the quenching/revival of the fluorescence emission. In an analytical context, the receptor is called, *sensor*, a word borrowed from the language of the macroscopic world: 'sensor, noun — from the Latin verb: *sentĭo, sentis, sensi, sensum, sentīre*, to perceive — a device that interacts with matter or energy and responds with a signal'. A glass electrode (interacts with matter) and a thermometer (with energy) are macroscopic sensors. In the molecular world, a sensor (a molecular or chemical sensor, sometimes called *chemosensor*)[91] typically responds to a variation of the concentration of the substrate (or *analyte*). The efficiency of the sensor is expressed by the ratio of the instrumental response Δx over the variation of the concentration of the analyte Δc. The higher the $\Delta x/\Delta c$ quotient, the more efficient the sensor. Most common chemical sensors are optical, operating either by light absorption (colorimetric sensors) or emission (fluorescent sensors). Their advantage lies in the fact that the properties investigated are visually perceived and, in addition, can be detected at very low concentration levels. Fluorescence, in particular, is a valued feature for its sensitivity and because the mechanisms for its control (quenching/revival) are known, a situation which can help the successful design of a sensor.

There exist several paradigms that govern the design of fluorescent sensors. The most popular is the *fluorophore–spacer–receptor* (FSR),[92] which requires that the receptor subunit and the fluorophore fragment are covalently linked by a spacer. Things must be arranged in such a way that the analyte–receptor interaction modifies the emission properties of the fluorophore. The FSR paradigm is pictorially illustrated in Figure 11.1.

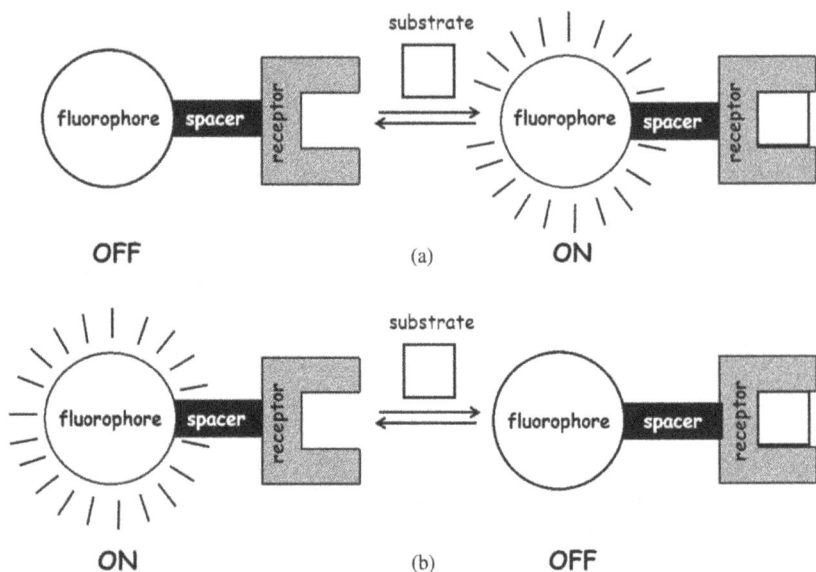

Figure 11.1. Fluorescence sensing of anions: the FSR paradigm; (a) OFF/ON type: the emission of the fluorophore is quenched through an intramolecular process, which is interrupted by the receptor–substrate interaction: recognition is signalled by a fluorescence revival; (b) ON/OFF type: the analyte itself quenches the fluorophore through an intra-complex process: recognition is signalled by quenching of fluorescence.

Two distinct mechanisms are possible: (a) an intra-molecular process involving receptor and fluorophore that quenches fluorescence pre-exists; the interaction of the substrate with the receptor interrupts the intra-molecular process and restores fluorescence (OFF/ON response); (b) in the absence of substrate, the fluorescence is on: the substrate, following the association with the receptor, quenches the proximate fluorophore through an intra-complex process (ON/OFF response).[93]

Figure 11.2 illustrates an example of type (a) behaviour, involving anthracene as a fluorophore, the cyclic tetramine cyclam as a receptor and the Zn^{II} ion as a substrate.

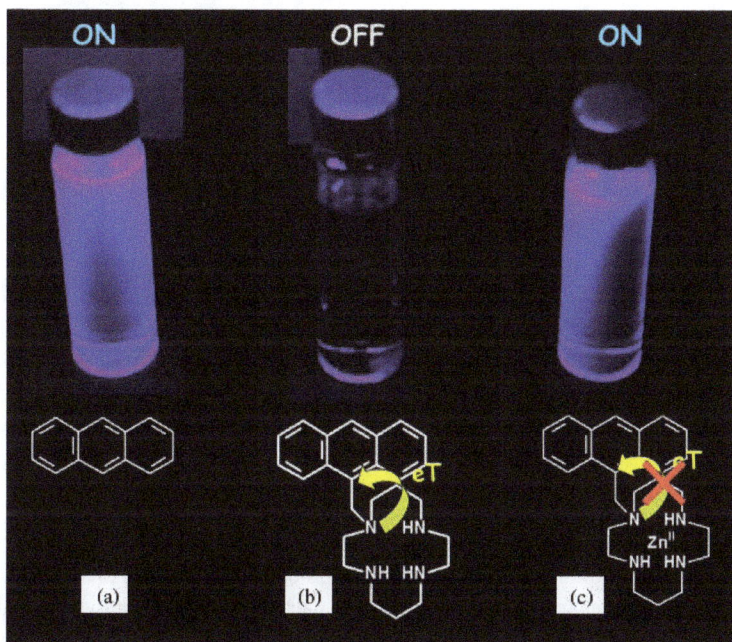

Figure 11.2. (a) A fluorescent sensor of Zn^{II} based on the FSR paradigm and displaying an OFF/ON response. Anthracene (An) and the cyclam–anthracene conjugate (cyAn) are dissolved in MeCN. An alone displays is blue fluorescence; (b) cyAn is not fluorescent, due to the occurrence on an eT process from N_{tert} to *An; (c) Zn^{II} engages N_{tert} in coordination and prevents eT, thus reviving fluorescence.

Anthracene is a classical fluorogenic molecule which, when excited at 355 nm, displays an emission spectrum with a well-defined vibrational structure with $\lambda_{max} = 405$ nm. An MeCN solution of anthracene emits a blue fluorescence as shown in Figure 11.2(a). The conjugate molecule in Figure 11.2(b) consists of the anthracene fluorophore and of the cyclic tetramine cyclam linked by a methylene group (the spacer). Cyclam is a good receptor for transition and post-transition metal ions (e.g. Zn^{II}). In the absence of the analyte (Zn^{II}), the fluorescence of the cyclam–anthracene conjugate is switched OFF due to the occurrence of a photoinduced electron transfer (eT) process. In particular, one electron of the lone pair of the tertiary amine nitrogen atom of the cyclam subunit is transferred to the excited anthracene subunit *An. Such an eT process quenches fluorescence. However, on coordination of Zn^{II}, the lone pair is engaged in the metal–amine bond and cannot longer be involved in the eT mechanism: anthracene fluorescence is switched ON (Figure 11.2(c)).

The dizinc(II) complex of the bistren cryptand **34** behaves as a fluorescent sensor for anions, providing an ON/OFF response (see Figure 11.3(a)).[94] Bistren cryptand **34** contains two 1,4-xylyl spacers and a 9,10-anthracenyl spacer, which operates as a fluorophore.

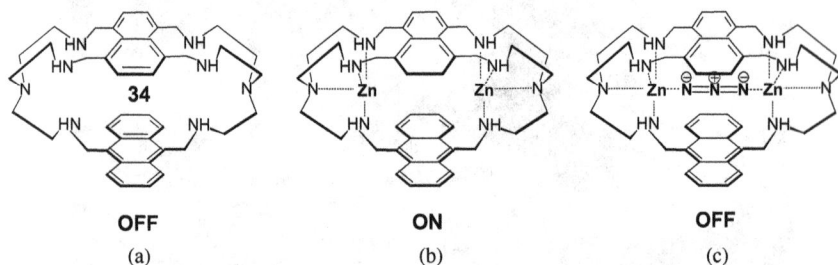

OFF	ON	OFF
(a)	(b)	(c)

Figure 11.3. An ON–OFF fluorescent sensor for the azide anion: (a) the bistren cryptand **34**, in an aqueous solution, does not fluoresce: emission is quenched due to the occurrence of an electron transfer (eT) process from a lone pair of one of the anthrylamine amine nitrogen atoms to the photoexcited anthracene moiety (OFF); (b) coordination of a Zn^{II} ion to each tren subunit engages nitrogen lone pairs and prevents the occurrence of an eT mechanism: the anthracene subunit can release its natural fluorescence (ON); (c) on inclusion of N_3^-, fluorescence is quenched again due to the transfer of an electron from a π bonding orbital of the anion to the half-filled π^* antibonding orbital of the excited anthracene fragment (OFF).[94]

The emitting behaviour in water of cryptand **34** and of its complexes, including anions or not, can be fully understood by carrying out spectrofluorimetric titration experiments at varying pH. In the first experiment, a solution containing **34** with excess acid was titrated with standard NaOH and the emission spectrum measured after each base addition. The spectro-fluorimetric titration profile, fluorescence intensity % ($I_F/I_0 \times 100$; I_F emission intensity at 400 nm, λ_{max}, at current pH; I_0 fluorescence intensity at 400 nm at pH = 0) vs pH, is shown in Figure 11.4(a) (circles).

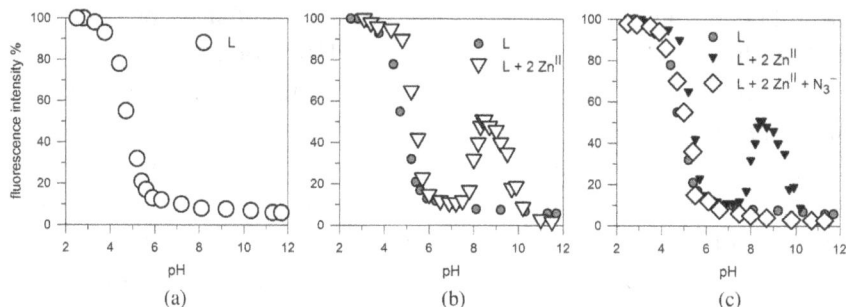

Figure 11.4. (a) Titration with standard NaOH of a solution of bistren cryptand **34** containing excess acid: following deprotonation of the two anthrylammonium groups of LH_6^{6+}, an eT process takes places from one amine nitrogen atom to the photexcited anthracene subunit, which quenches fluorescence; (b) the solution also contains 2 equivalent of Zn^{II}: at pH ~ 7, fluorescence is restored because the dimetallic cryptate $[Zn^{II}_2(\mathbf{34})(H_2O)_2]^{4+}$ forms and metal coordination stops the anthrylamine-to-*An eT process; at pH > 8, fluorescence is quenched again because one coordinated water molecule deprotonates and one electron is transferred from the metal-bound OH^- to *An; (c) the I_F vs pH profile superimposes on that of titration in (a): formation of the dimetallic cryptate is accompanied by the inclusion of an N_3^- anion: an electron is transferred from a π bonding orbital of azide to the half-filled π antibonding orbital of *An. In the vertical axis, $I_F/I_0 \times 100$ (I_F fluorescence intensity at 400 nm (λ_{max}) at current pH; I_0 fluorescence intensity at 400 at pH = 2).[94]

At pH = 2 and lower, where the cryptand is present in the hexa-protonated form, LH_6^{6+}, we observe full emission ($I_F/I_0 \times 100 = 100\%$). In the pH range 2–4, ammonium groups deprotonate and an eT process takes place from one of the proximate secondary anthrylamine groups to the excited anthracene subunit *An, which causes fluorescence quenching. I_F/I_0 decreases according to a sigmoidal profile and no emission is observed until the end of titration (pH = 12).

When the solution contains 2 equivalent of Zn^{II}, the I_F/I_0 vs pH profile is modified (see Figure 11.4(b)): fluorescence is restored at pH ~ 7, reaches its maximum intensity at pH = 8, then decreases again until complete quenching. Fluorescence revival at pH 7–8 has to be ascribed to the formation of the dimetallic cryptate $[Zn^{II}_2(L)(H_2O)_2]^{4+}$ (structural formula in Figure 11.3(b)), which interrupts the transfer of one electron of the lone pair of one anthrylamine nitrogen atom to a half-filled π antibonding molecular orbital of the excited anthracene subunit. Finally, the I_F/I_0 decrease after pH 8 is associated with the deprotonation of one of the metal-bound water molecules and to the occurrence of an eT process from the coordinated OH^- of the $[Zn^{II}_2(L)(H_2O)(OH)]^{3+}$ complex to *An.

The titration profile in Figure 11.4(c) refers to an aqueous solution as for the profile in Figure 11.4(b), but also containing 1 equivalent of NaN_3. The profile in Figure 11.4(c), superimposes well on that in Figure 11.4(a): in particular, N_3^- cancels the effect of fluorescence revival promoted by the coordination of the two Zn^{II} ions to the two tren subunits. Quenching is associated with the formation of the inclusion complex $[Zn^{II}_2(L)(N_3)]^{3+}$ (see structural formula in Figure 11.3(c)) and with the occurrence of an eT process from the electron rich N_3^- anion to *An. In particular, one electron is transferred from a filled π bonding orbital of azide to a half-filled π antibonding orbital of *An. Electron transfer is made possible by the steric constraints imposed by the cryptate framework, which places N_3^- and the anthracene moiety face-to-face. In particular, distances as short as 3 Å can be calculated through molecular modelling between the terminal nitrogen atoms of N_3^- and the closest carbon atoms of the anthracene fragment, a distance that allows a fast and efficient *through-space* eT process to occur.

Comparison of the $I_F/I_0 \times 100$ against pH profiles of Figures 11.4(b) and 11.4c indicates that at pH 8.0–8.5, the dizinc(II) cryptate could behave as an ON/OFF fluorescent sensor for N_3^-: in particular, azide inclusion into the $[Zn^{II}_2(\textbf{34})(H_2O)_2]^{4+}$ cryptate should be signalled by a substantial quenching of the anthracene fluorescence. Indeed, by titrating with NaN_3 an aqueous solution 10^{-4} M in **34** and 2×10^{-4} M in Zn^{II}, buffered to pH = 8.5 (called in the following 'receptor's solution'), a linear decrease of fluorescence was observed until the addition of 1 equivalent of azide (see Figure 11.5(a)). The curvature at 1 equivalent was smooth enough to allow a safe least-squares analysis of the titration profile, which gave a value of log $K = 5.8 \pm 0.1$ (conditional constant at pH = 8.5). Then, the receptor's solution was titrated with a series of anions: NO_3^-, HCO_3^-, SO_4^{2-}, Cl^-, Br^-. In all cases, no spectral modification, and no I_F decrease were

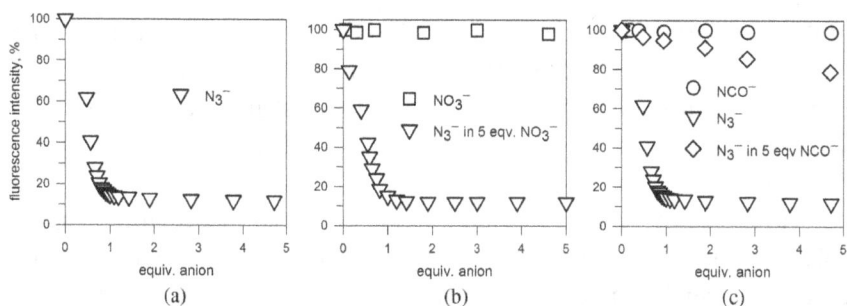

Figure 11.5. Spectrofluorimetric titration profiles of (a) a solution 10^{-4} M in **34** and 2×10^{-4} M in Zn^{II}, buffered to pH = 8.5 (thus containing the $[Zn^{II}_2(L)(H_2O)_2]^{4+}$ cryptate (L = **34**), 'receptor's solution') titrated with NaN_3; (b) squares, titration with $NaNO_3$ (no modification of the emission spectrum; the same behaviour observed on titration with HCO_3^-, SO_4^{2-}, Cl^-, Br^-); triangles, titration with N_3^- of the 'receptor's solution' also containing 5 equivalent of one of the anions NO_3^-, HCO_3^-, SO_4^{2-}, Cl^-, Br^-; (c) circles, titration with $NaNCO$; diamonds, titration with NaN_3 of the 'receptor's solution' also containing 5 equivalent of NCO^-; triangles, same profile as in (a), reported for comparative purposes.[94]

observed even after the addition of 5 equivalent anion (Figure 11.5(b)). Moreover, titration with NaN_3 of the receptor's solution also containing 5 equivalent of one of the above-mentioned anions produced the same profile observed in the absence of anions (see Figure 11.5(b)). This indicates that the considered anions do not compete with N_3^- for inclusion, implying in particular that corresponding K values should be $\leq 10^{3.8}$ (i.e. at least 100-fold lower than K for N_3^-).

The case of NCO^- is unique. On addition of NCO^- to the receptor's solution, no modification of the emission spectrum was observed. However, the profile obtained on titration with N_3^- was remarkably affected by the presence of NCO^- in the receptor's solution: the greater the concentration of NCO^-, the less steep the $I_F/I_0 \times 100$ decrease, indicating severe competition of the two anions for inclusion into the cryptate. The profile reported in Figure 11.5(c), diamonds, refers to a solution containing 5 equivalent of NCO^-. From competitive titration experiments, a log K of 6.5 ± 0.1 could be calculated for NCO^-. Thus, NCO^- shows a 5-fold greater affinity for $[Zn^{II}_2(L)(H_2O)_2]^{4+}$ than N_3^-, but, due to its negligible reducing tendencies, when included in the cryptate it is unable transfer an electron to the nearby *An fragment, thus failing to signal the occurrence of the recognition process. N_3^- and NCO$^-$ have a comparable bite (2.34 and 2.42 Å, respectively), and the rather high values of K reflect the favourable matching of the anion bite length with the distance between the two vacant axial positions of the two Zn^{II} centres (in the present case, slightly favourable to NCO$^-$).

Titration experiments showed that the other rod-like triatomic anion, NCS$^-$, quenches *An fluorescence, but according to a much less steep profile, to which a much lower value of log K corresponds: 2.45 ± 0.05 (a feature already observed with the dicopper(II) cryptate of **28**). NCS$^-$ is a one-electron reducing agent with strength comparable to that of N_3^- (NCS·/NCS$^-$ potential: 1.62 V vs NHE; $N_3 \cdot / N_3^-$: 1.33), which accounts for the occurrence of an intra-complex photoinduced eT process and fluorescence quenching. However, the much greater bite length (2.75 Å) induces a severe endergonic rearrangement of the cryptate framework, making NCS$^-$ inclusion 2200 times less favourable than for N_3^- (and 11,200 times than for NCO$^-$).

12

Anion Fluorescence Sensing by Dimetallic Bistren Cryptates: The Indicator Displacement Paradigm

The FSR approach to the design of fluorescent sensors for anions is not always convenient and satisfying for many reasons: first, the covalent linking of the receptor and fluorophore is not straightforward and may require a tedious multistep synthesis. For instance, in the case of bistren-based receptors like **34**, which contains spacers of different nature, the 3 + 2 Schiff base condensation procedure cannot be employed. Moreover, the mechanism of the modification of the fluorescent emission is often unpredictable and cannot be planned with certainty. As an example, the $[Zn^{II}_2(L)(H_2O)_2]^{4+}$ sensor (L = **34**) considered in Chapter 11 responds to N_3^-, which brings with itself the 'signal transduction mechanism' (in this case, the capability to transfer an electron to the photoexcited anthracene subunit, thus quenching fluorescence), but does not respond to NCO⁻, which on its own forms a very stable inclusion complex, but is reluctant to release an electron and leaves fluorescence emission undisturbed.

Moreover, if the receptor must contain a metal, the choice is limited to those showing a d^{10} electronic configuration and not displaying redox activity (e.g. Zn^{II}), which excludes the numerous and hardened troops of transition metal ions. But there is another aspect which makes the FSR not too attractive. In most cases, FSR fluorescent sensors operate through an ON/OFF mechanism, like $[Zn^{II}_2(L)(H_2O)_2]^{4+}$ (L = **34**). Among people using fluorescence as a signal, the ON/OFF response is considered less valuable than the OFF/ON response. The reason was explained to me by a friend and colleague, Fabio Grohovaz, cell physiologist at San Raffaele Hospital, Milan, according to whom 'it is easier and more comfortable to detect an electrical torch switching on in the Black Forest, than the light of an apartment switching off in the night in Manhattan'. The aphorism is pictorially illustrated in Figure 12.1.

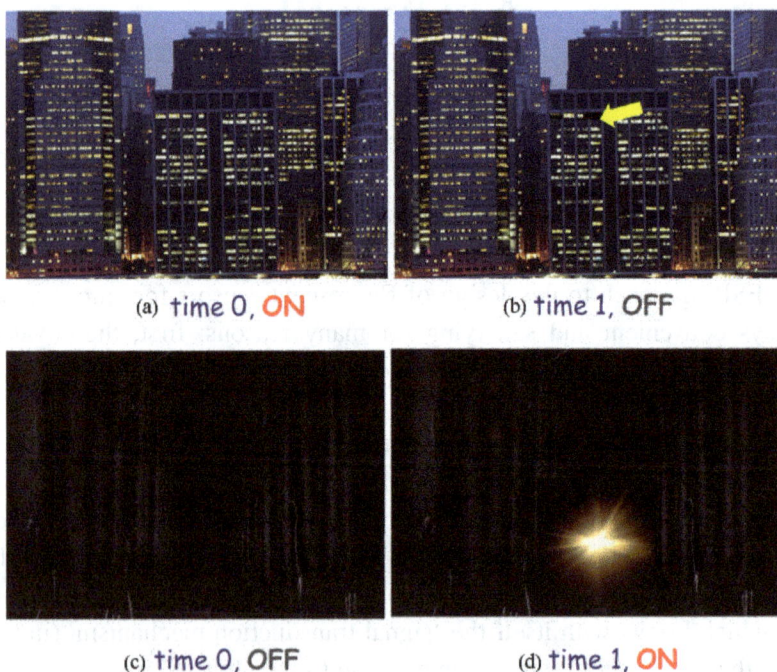

(a) time 0, ON   (b) time 1, OFF

(c) time 0, OFF   (d) time 1, ON

Figure 12.1. (a) and (b) The light of the apartment is switched off (see the yellow arrow in (b)); (d) and (c) the electrical torch is switched on.

This sentence reflects the point of view of a scientist used to studying through a fluorescence microscope the behaviour under stimulus of a cell in which a fluorescent indicator had been injected using a microsyringe.

Fortunately, there exists another sensing approach that remedies most of the drawbacks of the FSR paradigm: this approach makes use of an optical indicator, **In**, either coloured or fluorescent.[95,96] In particular, **In** is bound, through a non-covalent interaction, to the receptor **R**, to give a kinetically labile [**R**···**In**] complex. Then, the envisaged substrate **S**, displaying a special affinity for **R**, is added to the solution: **S** displaces from the receptor cavity the loosely bound **In**, which is released to the solution, according to the fast exchange equilibrium (12.1):

$$[\mathbf{R}\cdots\mathbf{In}] + \mathbf{S} \leftrightarrows [\mathbf{R}\cdots\mathbf{S}] + \mathbf{In} \tag{12.1}$$

For signalling purposes, it is required that an optical property of the indicator (e.g. colour) distinctly changes depending whether **In** is bound to **R** or is free. Under these circumstances, substrate recognition, and indicator release, will be signalled by a sharp colour change.

The paradigm of the *indicator displacement* can be conveniently utilised for fluorescent sensing of anions by *metal-containing receptors*. First, the indicator should be a fluorogenic fragment and should be also able to coordinate the metal centre of the receptor: this goal can be achieved by using a –COO⁻ containing fluorophore. In fact, the –COO⁻ group displays pronounced coordinating tendencies towards transition metal ions. Then, things should be arranged in such a way that the fluorescent emission changes drastically following the displacement of the indicator. In particular, for profiting from the valuable OFF/ON signalling, (i) the fluorophore must be quenched when bound to the metal and (ii) it must display its full fluorescence when released to the solution. In other words, the metal-containing receptor must be able to quench the fluorescent indicator. To do that, a photophysically active ('guilty') metal should be chosen, instead of the photophysically inactive ('innocent') Zn^{II} centre used in the FSR paradigm. This is the case of Cu^{II}, which, due to the incomplete filling of the $3d$ level and to the significant redox activity ($Cu^{I} \leftarrow Cu^{II} \rightarrow Cu^{III}$), is able to quench any nearby fluorophore through either an ET or an eT process, respectively. As an extra benefit, Cu^{II} is highest in the Irving–Williams series of metal ions and gives very stable complexes with polyamines, including octamine bistren cryptands.

The application of the indicator displacement paradigm to the fluorescence sensing of polyatomic anions by using cryptand **28** is discussed below as a working example.[97]

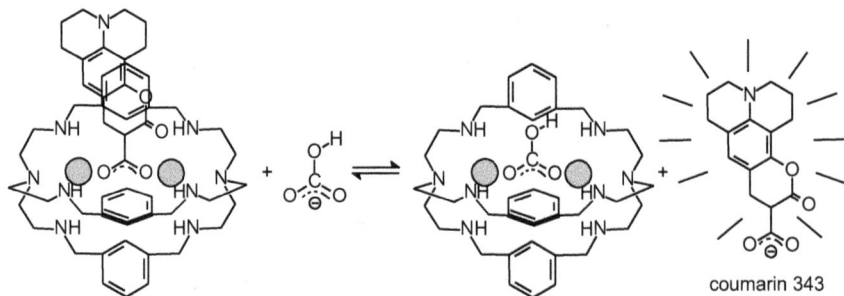

coumarin 343

Figure 12.2. Fluorescence sensing of HCO_3^- by the indicator displacement paradigm. When complexed by the dicopper(II) cryptate, the Coumarin 343 dye is non-fluorescent. When displaced from the cryptate by the incoming HCO_3^- anion, the indicator displays its natural intense fluorescence.[97]

First, as a fluorescent indicator Coumarin 343 is chosen (see structural formula in Figure 12.2), which possesses a carboxylate function capable of bridging the two Cu^{II} centres of the cryptate and is strongly fluorescent in its anionic form. On titrating a solution of 10^{-7} M of Coumarin 343, buffered to pH 7 (HEPES 0.05 M), with a solution containing 1 equivalent of cryptand **28** and 2 equivalent of Cu^{II}, complete quenching of the dye's emission is observed, as shown by the titration profile in Figure 12.3(a). Curve fitting of the titration profile (see Figure 12.3(a)), through a nonlinear least-squares procedure, was consistent with the formation of a complex of 1:1 stoichiometry, whose association equilibrium constant, K (conditional value at pH 7), was 4.8 ± 0.1 log units. Fluorescence quenching has to be ascribed to an intra-complex electronic energy transfer (ET) process involving the photoexcited Coumarin 343 fragment and the Cu^{II} centre(s).

Based on these premises, an aqueous solution 2×10^{-4} M in $[Cu^{II}_2(L)(H_2O)(OH)]^{3+}$ and 10^{-7} M in Coumarin 343, buffered to pH = 7 with HEPES, was titrated with a variety of anions for fluorescent sensing

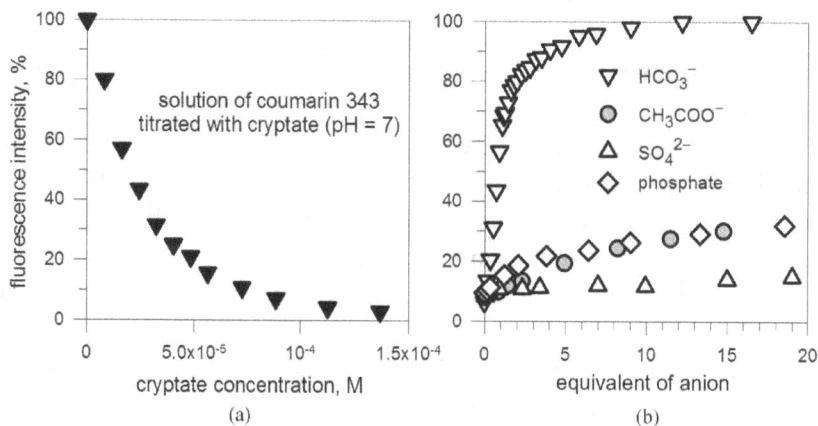

Figure 12.3. (a) Titration of a 10^{-7} M solution of the fluorescent indicator Coumarin 343 ($\lambda_{exc} = 424$ nm, λ_{em} 487 nm) with the dimetallic cryptate $[Cu^{II}{}_2(L)]^{4+}$ (L = **28**) in an aqueous solution buffered with HEPES to pH 7 (at this pH the cryptate is present in the form $[Cu^{II}{}_2(L)(H_2O)(OH)]^{3+}$); a 1:1 inclusion complex forms, whose association constant is 4.8 ± 0.1 log units. Quenching of the dye is ascribed to the occurrence of an intra-complex electronic energy transfer (ET) process, which involves the excited fluorophore and the Cu^{II} centre(s) inside the dimetallic complex; (b) profiles (emission intensity at 487 nm against equivalent of anion) obtained over the course of competitive titrations of an aqueous solution 2×10^{-4} M in $[Cu^{II}{}_2(\mathbf{28})(H_2O)(OH)]^{3+}$ and 10^{-7} M in Coumarin 343, buffered to pH = 7 with HEPES, with standard solutions of selected anions. HCO_3^- is able to displace from the dicopper(II) cryptate the indicator, which is released to the solution and displays its full emission. Other anions, for example, phosphate, CH_3COO^-, SO_4^{2-} do not compete successfully with Coumarin 343 for the cryptate and induce only a slight fluorescence enhancement.[97]

purposes. The combination of receptor and indicator in the chosen molar ratio will be called in the following 'chemosensing ensemble', according to the definition by Eric Anslyn, University of Texas at Austin.[98]

Titration of the chemosensing ensemble solution with a standard solution of carbonate (see the fluorescence intensity against anion equivalents profile in Figure 12.3(b), triangles down) resulted in an almost complete recovery of the fluorescence emission, which indicated successful competitive binding of HCO_3^- and displacement of the indicator from the host cavity. Full release of Coumarin 343 emission was also observed when the 'chemosensing ensemble' solution was titrated with N_3^- and NCO^- anions. On the other hand, titration with other ambidentate anions (NCS^-, NO_3^-,

SO_4^{2-}, $H_2PO_4^-/HPO_4^{2-}$, $HCOO^-$, CH_3COO^-) caused only a slight fluorescence enhancement (see in Figure 12.3(b) some selected titration profiles).

The selectivity pattern discussed above can be accounted for by assuming that the conditional constants for the inclusion of HCO_3^-, N_3^- and NCO^- into the dicopper(II) cryptate are significantly larger than that observed for the coumarin indicator (log $K = 4.8$), whereas all the other investigated anions must show values of log K substantially lower than 4.8.

In order to confirm this hypothesis, the equilibrium constants for the inclusion of the investigated anions in $[Cu^{II}{}_2(L)(H_2O)(OH)]^{3+}$, in a solution buffered to pH 7, were determined through spectrophotometric titration experiments, in particular considering the development of the charge transfer absorption bands mainly in the UV region, according to the procedure described in Chapter 10 for the same dicopper(II) cryptate. The difference is that in this case log K values have been determined at pH = 7 (HEPES buffer), whereas in the previous case they had been determined at pH = 8 (lutidine buffer). The plot of log K values against anion bite length shown in Figure 12.4(a) (pH = 7) is quite similar to that observed at pH = 8 and shown in Figure 10.4(b).

In Figure 12.4(a) a dashed horizontal line has been drawn at log $K = 4.8$, which corresponds to the constant of the inclusion equilibrium of Coumarin 343. The figure clearly explains how the indicator displacement paradigm works: (i) anions lying above the horizontal line displace the indicator from the cryptate cavity and are visually and spectrofluorimetrically sensed; (ii) anions lying below the horizontal line do not displace the metal-bound indicator, and so fluorescence remains quenched.

Among investigated anions, carbonate is the most important from the point of view of analytical applications. In this regard, a calibration curve was obtained over the concentration range 2×10^{-5} M $\leftrightarrow 2 \times 10^{-4}$ M, which is shown in 12.4b. Such a calibration curve was then used for the quantitative determination of carbonate in samples of commercial mineral waters. In a typical experiment, a small amount of the sample (10–500 μL) was added to the chemosensing ensemble solution (10 mL, 2×10^{-4} in cryptate and 10^{-7} M in Coumarin 343, HEPES buffer pH 7). Fluorescence intensity was measured, and the amount of carbonate was obtained by

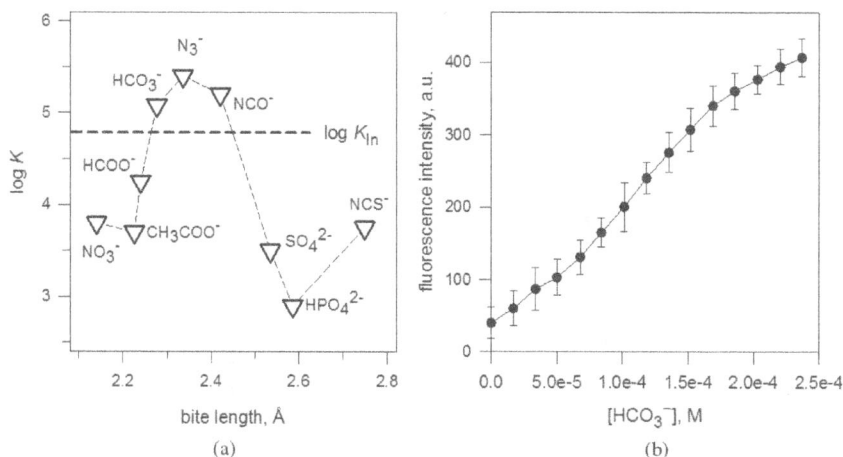

Figure 12.4. (a) Selectivity pattern for the inclusion of poly-atomic anions into the dimetallic receptor $[Cu^{II}_2(L)(H_2O)(OH)]^{3+}$ at pH 7 (L = **28**). Conditional constants K of the anion inclusion equilibria were determined through spectrophotometric titrations. The dashed horizontal line corresponds to the log K value for the formation of the receptor/indicator complex (4.8). Only the anions whose log K value lies above the horizontal line are able to displace the indicator from the cryptate receptor and can be detected spectrofluorimetrically; (b) calibration curve for the determination of the concentration of HCO_3^- in aqueous solution by measuring the fluorescence intensity of a solution containing the chemosensing ensemble $[Cu^{II}_2(L)]^{4+}$/Coumarin 343, buffered to pH = 7.[97]

interpolation on the calibration curve in Figure 12.4(b). The interval of the calibration curve includes the concentration of all the commercially available mineral waters. The fluorimetric procedure seems more convenient and more selective than that legally required (in Italy) to determine HCO_3^- in mineral water, which is based on the volumetric titration of the alkalinity of the solution with standard hydrochloric or sulphuric acid: in fact, the latter procedure determines the total concentration of basic substances present in solution and does not discriminate carbonate from other bases.

13

Nucleotide Recognition and Sensing by a Dicopper(II) Bistren Cryptate

Nucleotides are the building blocks of nucleic acids (RNA and DNA) and are composed of a nitrogenous base (or nucleobase: adenine, guanine, citosine, thymine, uracyl), a five-carbon sugar (ribose or deoxyribose) and at least one phosphate group. The nucleobase along with the sugar gives the nucleoside, the nucleoside along with phosphate(s) gives the nucleotide. Nucleoside polyphosphates (NPPs) play a central role in many biological processes. In particular, they provide a universal source of chemical energy (adenosine triphosphate (ATP) and guanosine triphosphate),[99] participate in cellular signalling (cyclic guanosine monophosphate (GMP) and cyclic adenosine monophosphate (AMP))[100] and are part of important cofactors of enzymatic reactions (coenzyme A, flavin adenine dinucleotide, flavin mononucleotide, and nicotinamide adenine dinucleotide phosphate).[101]

Typically, nucleotides are recognised by receptors capable of establishing electrostatic and/or hydrogen bonding interactions with the phosphonate groups.[102] It was considered that the two copper(II) centres of a bistren cryptate could establish two-point interactions with the nucleotide, providing a more energetic and more selective way of recognition.[103] The following nucleotides (nucleoside monophosphates, NMPs) were considered for recognition and sensing through the formation of a bistren cryptate.

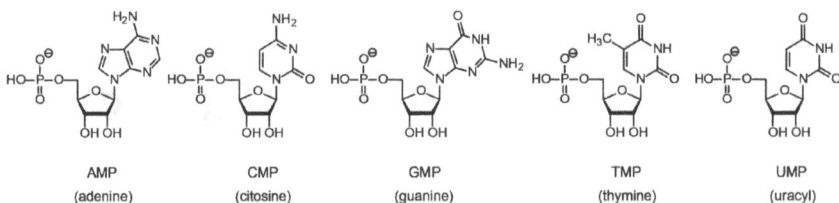

| AMP | CMP | GMP | TMP | UMP |
| (adenine) | (citosine) | (guanine) | (thymine) | (uracyl) |

Each NMP anion contains three potentially coordinating protruding groups, suitable for coordination of the two Cu^{II} centres of a bistren cryptate of the appropriate cavity size: (i) the phosphate ion (donor: a phosphate oxygen atom, formally detaining ½ negative charge); (ii) an aniline group (neutral, sp^3 nitrogen atom, only AMP, CMP and GMP); (iii) a carbonyl fragment (neutral, sp^2 oxygen atom, AMP).

In order to generate a bistren cryptate of size suitable for the satisfactory inclusion of monophosphate nucleosides, 2,2-bis(5-formyl-2-furyl)-propane (**35**) was chosen as a dialdehyde for the Schiff base condensation with tren, according to the 3:2 molar ratio. The corresponding hexaimine cryptand was obtained in good yield and was made 'irreversible' through the hydrogenation of the six C=N bonds, to give **36**.

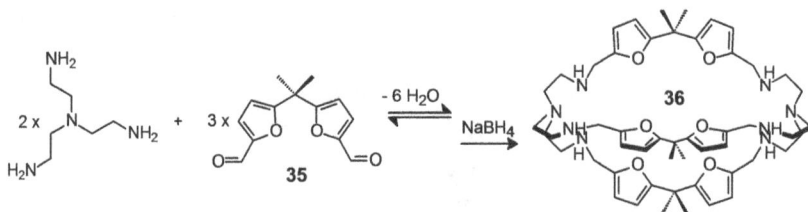

Equilibrium studies on the formation of the dicopper(II) cryptate complex were carried out in a water/methanol solution (50:50, v/v). Such a composition ensured full solubility of the investigated systems as well as a safe control of the instrumental determination of pH. Pertinent equilibrium constants (protonation of the cryptand and formation of the cryptate complexes) were determined through potentiometric titration experiments. In particular, Figure 13.1(a) shows the distribution diagram of the species present at the equilibrium in the 2–10 pH interval for a solution of 5×10^{-4} M in cryptand **36** and 10^{-3} M in $Cu^{II}(CF_3SO_3)_2$. In particular, at low pH values, the following species form: $[Cu^{II}(LH_3)]^{5+}$, $[Cu^{II}(LH_2)]^{4+}$ and $[Cu^{II}(LH)]^{3+}$, in which one tren subunit binds one metal and the other is protonated. The dimetallic complex $[Cu^{II}_2(L)]^{4+}$ begins to form at pH 3 and reaches its maximum concentration (85%) at pH 6. On increasing pH, the two metal-bound water molecules deprotonate stepwise and the hydroxo species $[Cu^{II}_2(L)(OH)]^{3+}$ and $[Cu^{II}_2(L)(OH)_2]^{2+}$ form. The distribution diagram is shown in Figure 13.1(a).

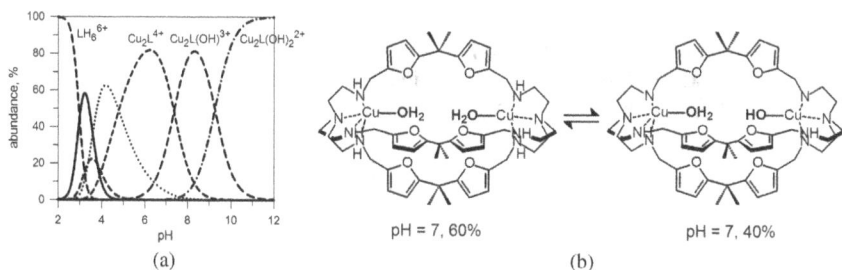

Figure 13.1. (a) Percent concentration profiles of the species present at the equilibrium over the course of the titration with standard NaOH of a solution 5×10^{-4} M in cryptand **36** and 10^{-3} M in $Cu^{II}(CF_3SO_3)_2$, containing excess acid. At pH = 7, the two complex species $[Cu^{II}_2(L)(H_2O)_2]^{4+}$ (60%) and $[Cu^{II}_2(L)(H_2O)(OH)]^{3+}$ (40%) coexist at the equilibrium. Their hypothesised structural arrangements are sketched in (b).[103]

At pH = 7, the two complex species $[Cu^{II}_2(L)(H_2O)_2]^{4+}$ (60%) and $[Cu^{II}_2(L)(H_2O)(OH)]^{3+}$ (40%), whose hypothesised structural arrangements are sketched in Figure 13.1(b), coexist at equilibrium. As a fluorescent indicator used in the preparation of the 'chemosensing ensemble' for NMP recognition, 6-carboxyfluorescein was chosen, whose absorption band is centred at 495 nm and the emission band shows a maximum at 517 nm. At pH 7, the indicator detains a triple negative charge, and it is hypothesised that it offers as donor atoms a carboxylate oxygen atom and a phenolate oxygen atom (indicated in the structural formula in Figure

13.2 with stars). In fact, the bite of these two oxygens is expected to encompass the two copper(II) centres of the cryptate, as judged from molecular modelling.

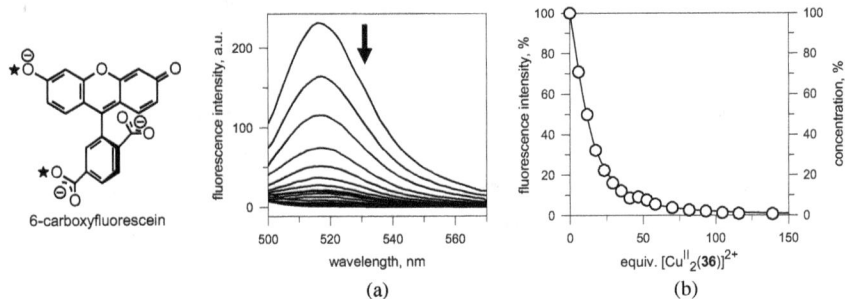

Figure 13.2. (a) Emission spectra recorded over the course of the titration of a solution of an MeOH/H$_2$O (50:50) 2×10^{-7} M in the indicator 6-carboxyfluorescein, buffered to pH 7 with HEPES, with a solution 2.3×10^{-3} M in [Cu$^{II}_2$(L)]$^{4+}$ (L = **36**, λ_{exc} = 492 nm); (b) symbols, left axis: normalised intensity of the emission at 516 nm ($I/I_0 \times 100$, I_0 = emission intensity of the indicator alone); line, right axis: percent concentration of the uncomplexed indicator.[103]

Figure 13.2(a) displays the family of emission spectra recorded over the course of the titration of a solution of MeOH/H$_2$O (50/50) 2×10^{-7} M in the indicator, buffered to pH 7 with HEPES buffer, with a solution 2.3×10^{-3} M in the receptor (which consists of a mixture of [Cu$^{II}_2$(L)(H$_2$O)$_2$]$^{4+}$ and [Cu$^{II}_2$(L)(H$_2$O)(OH)]$^{3+}$). On cryptate addition, the fluorescence intensity of the dye is progressively quenched, demonstrating the occurrence of an interaction with the metal centre(s) of the receptor. The extent of fluorescence quenching is illustrated by the titration profile (see symbols in Figure 13.2(b)). Complete quenching of fluorescence takes place on addition of more than 100 equivalent of the receptor, a consequence of the high dilution of the solution.

Based on this, the 'chemosensing ensemble' solution was prepared (50:50 MeOH/H$_2$O (v/v), 2×10^{-5} M in [Cu$^{II}_2$(L)]$^{4+}$ (L = **36**) and 2×10^{-7} M in 6-carboxyfluorescein, adjusted to pH 7 with HEPES buffer) and was titrated with a standard solution of the chosen NMP. Figure 13.3(a) shows the family of spectra recorded over the course of the titration. The dashed

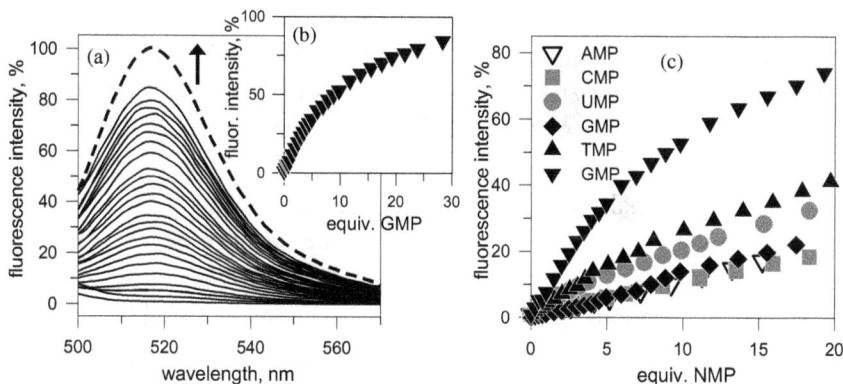

Figure 13.3. (a) Emission spectra taken during the titration of a 50:50 (v/v) MeOH/H$_2$O solution 2 × 10^{-5} M in [Cu$^{II}_2$(L)]$^{4+}$ (L = **36**) and 2 × 10^{-7} M in 6-carboxyfluorescein, buffered to pH 7 with HEPES, with a solution of GMP (λ_{exc} = 492 nm); dashed line refers to the spectrum of the uncomplexed indicator, under the same conditions; the emission intensity has been normalised with respect to the emission of the free indicator; (b) normalised emission intensity at 516 nm; (c) titration profiles obtained for titrations with different nucleoside monophosphates (NMPs) of the same 'chemosensing ensemble' solution, as described for GMP. In all titrations the emission intensity at 516 nm has been normalised with respect to the emission of the uncomplexed indicator.[103]

line refers to the emission spectrum of 6-carboxyfluorescein alone taken under the same conditions.

Figure 13.3(b) displays the titration profile based on the normalised emission intensity at 516 nm: it is observed that on addition of 30 equivalent of GMP the emission spectrum of the fluorescent indicator has reached ca. 80% of its limiting value. This means that 80% of the indicator has been displaced from the cryptate complex. Similar titration experiments were carried out with the other NMPs, and corresponding titration profiles at 516 nm are shown in Figure 13.3(c). In all cases, the amount of displaced indicator after excess addition of NMP was markedly lower than in the case of GMP, for which receptor [Cu$^{II}_2$(L)]$^{4+}$ (L = **36**) exerts a significant selectivity. In particular, through computer analysis of the titration profiles, the formation constants of the cryptate/NMP complexes at pH 7 were calculated, which gave the following sequence of stability (log K values in parentheses)

GMP (4.7) >> TMP (4.2) > UMP (4.0) > CMP (3.7) = AMP (3.7)

The stability of the cryptate/nucleotide complexes depends upon two main factors (i) the nature of NMP donor atoms coordinated to the Cu^{II} centres of the receptor and (ii) the capability of NMP donor atoms to fit the Cu^{II}---Cu^{II} distance. As far as point (i) is concerned, it is reasonable to assume that one of the donor atoms is an oxygen of the phosphonate group, which formally detains half of the negative charge. On the other hand, no atom of the nucleobase possesses a negative charge, a circumstance which makes its donor tendencies quite low. However, some of the considered nucleotides, GMP, TMP and UMP, contain an acidic secondary amide group that, under the conditions of the investigation, may undergo deprotonation and transfer the negative charge to the adjacent oxygen atom. Figure 13.4 illustrates such a mechanism for UMP.

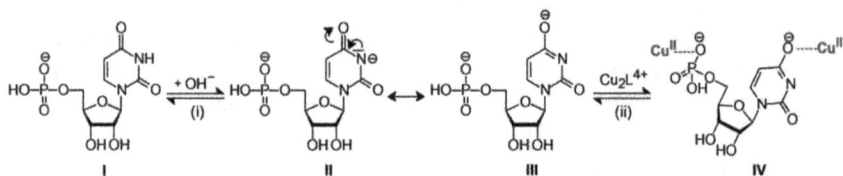

Figure 13.4. Binding of UMP to the two copper centres of the $[Cu^{II}_2(36)]^{4+}$ cryptate. On deprotonation of the amide N–H fragment of **I**, the negative charge moves through a π mechanism from the nitrogen atom (keto form **II**) to the oxygen atom (enolate form **III**). The enolate electronic arrangement is stabilised by the coordination to one Cu^{II} centre of the cryptate, and the cryptate/UMP complex **IV** is formed. A similar binding mechanism is suggested for the other NMP containing a secondary amide group (GMP and TMP).

The secondary amide group of the nucleobase of **I** is poorly acidic, and so the dissociation equilibrium (i) is displaced to left. In the deprotonated form **II**, scarcely present in solution, the negative charge moves through a π mechanism from the nitrogen atom (keto form **II**) to the oxygen atom (enolate form **III**). The enolate electronic arrangement is strongly stabilised by coordination to one metal centre of the receptor. Thus, the interaction of UMP with the cryptate (as well as that of GMP and TMP) profits from the coordination to the two Cu^{II} centres by two formally negative oxygen atoms (from phosphonate and enolate). Such a mechanism is precluded to AMP (whose adenine base possesses a tertiary amide group) and CMP (whose cytosine base does not possess amide groups at all). This accounts for the lowest values of the stability constants observed for AMP and CMP, forced to bind one metal centre with the neutral and *per se* poorly coordinating aniline nitrogen atom.

The different stability of the cryptate complexes of secondary amide-containing NMPs can be tentatively accounted for on a geometrical basis. Figure 13.5(a) displays the crystal structures of the three nucleotides with the distances between the suggested donor atoms.

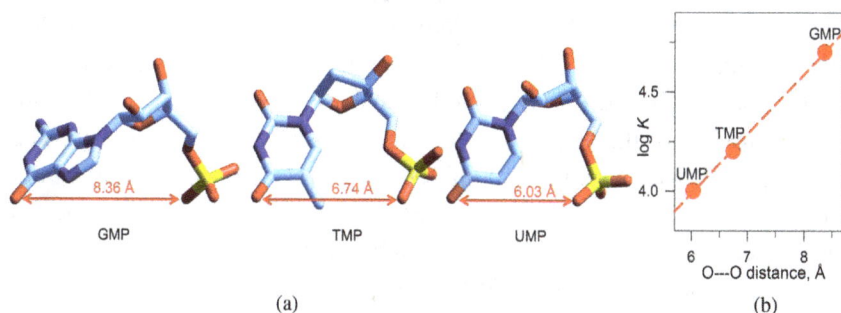

Figure 13.5.   (a) The crystal structures of nucleoside monophosphates GMP,[104] TMP[105] and UMP.[106] Hydrogen atoms have been omitted for clarity. Double-head arrows indicate the most favourable distances between the oxygen donor atoms which can bind the two metal centres of a dicopper(II) cryptate. One oxygen atom belongs to the phosphonate subunit, the other to the carbonyl oxygen of the nucleobase. The latter oxygen, under the condition of the investigation — pH = 7, interaction with the $[Cu^{II}_2(L)]^{4+}$ cryptate (L = **36**) — is supposed to have a dominant phenolate nature; (b) a linear relationship between the distance of the two above mentioned oxygens and the log K values for the formation of the NMP/cryptate complex at pH = 7.

In Figure 13.5(a), the distances between the potentially coordinating oxygen atoms (one from phosphonate, one from the carbonyl fragment of the secondary amide group of the nucleobase) are indicated with a double-head arrow. It is assumed that O---O distances do not change following amide –N–H deprotonation and electronic rearrangement to phenolate. Quite interestingly, the sequence of O---O distances (the 'bite' of NMPs) parallels the sequence of the solution stabilities: UMP < TMP < GMP. Even, as shown in the diagram in Figure 13.5(b), the two quantities are linearly correlated (indeed, putting three points on a straight line is not a difficult matter!). It has to be recalled that the two partially negative oxygens of the NMP replace the water/hydroxide oxygens of the $[Cu^{II}_2(L)(H_2O)_2]^{4+}$ and $[Cu^{II}_2(L)(H_2O)(OH)]^{3+}$ forms of the receptor. It follows that the ideal O---O distance of receptor must be ≥ 8.5 Å. Nucleotides characterised by a shorter bite (TMP and UMP) force the receptor to the contract its major ellipsoidal axis, to the detriment of the stability of the complex.

In the discussion on the binding properties of dimetallic cryptates, terms like 'inclusion' and 'incorporation' are of current use. Studies on the complexation equilibria involving $[Cu^{II}_2(L)]^{4+}$ (L = **28**) and small inorganic anions (N_3^-, HCO_3^-) have demonstrated the correctness of such a lexical choice. However, this does not seem appropriate to describe the interaction of $[Cu^{II}_2(L)(H_2O)_2]^{4+}$ and $[Cu^{II}_2L(H_2O)(OH)]^{3+}$ (L = **36**) with NMPs, in view of the size and of the complex shape of the substrate. The exact geometrical arrangement of the cryptate/NMP complexes is not known in the absence of crystallographic structural data. However, a rough idea of the coordination mode can be speculated from the calculated structure of the $[Cu^{II}_2(L)(GMP)]^{2+}$ complex shown in Figure 13.6.

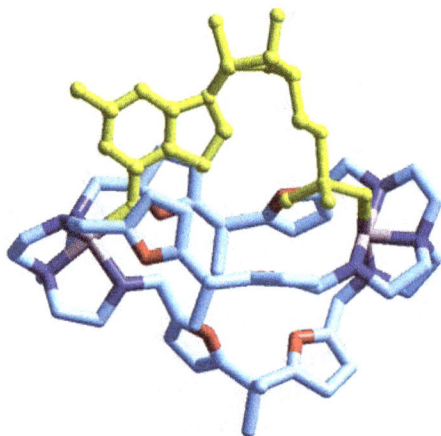

Figure 13.6. The calculated (MM+) structure of the complex $[Cu^{II}_2(L)(GMP)]^{2+}$ (L = **36**). GMP (yellow ball and stick representation) retains a doubly negative charge due to the deprotonation of the secondary amide group of the guanosine nucleobase and to the phosphonate group. One phosphonate oxygen and one enolate oxygen coordinate the two Cu^{II} centres of the cryptate.

The structure shows that the double negative GMP anion is well outside of the cryptate cavity. The cryptate exposes its two copper(II) centres towards the phosphonate and enolate oxygen atoms of the nucleotide. Thus, rather than a *cryptate effect*, we are have in the presence a *reinforced chelate effect*. In particular, the rigidity of the macrobicyclic dimetallic receptor enhances its binding tendencies towards a spatially fixed couple of negatively charged oxygen atoms.

It has been shown that the phosphonate group plays a key role in the interaction of nucleotides with the $[Cu^{II}_2(L)]^{4+}$ receptor (L = **36**). Thus, it seemed useful to verify the binding tendencies of mono-, di- and tri-phosphates towards the same cryptate under the same experimental conditions (50:50 MeOH/H$_2$O solution, pH buffered to 7 with HEPES), using 6-carboxy-fluorescein as a fluorescent indicator. In particular, the 'chemosensing ensemble' solution (2×10^{-5} M in $[Cu^{II}_2(L)]^{4+}$ (L = **36**) and 2×10^{-7} M in 6-carboxy-fluorescein, adjusted to pH 7 with HEPES buffer) was titrated with a solution of sodium phosphate (mono-, di- and tri-). Corresponding titration profiles are shown in Figure 13.7.

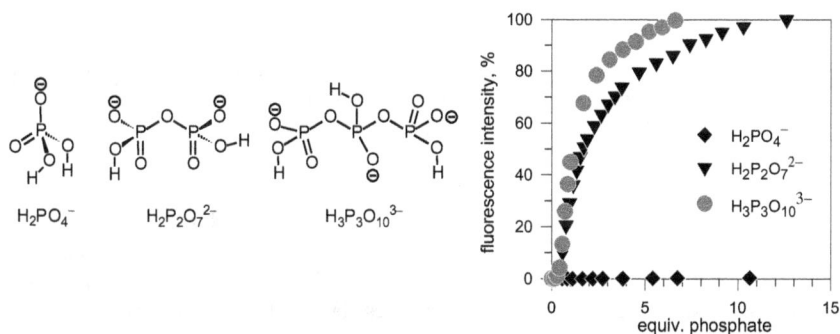

Figure 13.7. Spectrofluorimetric titration profiles obtained for different phosphates by monitoring the displacement of the indicator 6-carboxy-fluorescein from the receptor $[Cu^{II}_2(L)]^{4+}$ (L = **36**); the 'chemosensing ensemble' solution (50:50 MeOH/H$_2$O 2×10^{-5} M in $[Cu^{II}_2(L)]^{4+}$, 2×10^{-7} M in 6-carboxy-fluorescein, adjusted to pH 7 with HEPES buffer) was titrated with a solution of the chosen phosphate. The emission intensity at 516 nm has been normalised with respect to the emission of the uncomplexed indicator.[103]

It is observed that monophosphate fails to displace the indicator from the cryptate. Surely, its bite is too short to encompass the Cu^{II}---Cu^{II} distance and the inclusion of two monophosphates may be sterically disfavoured (even if it has been observed in some cases that two monophosphate anions may establish intermolecular H-bond interactions to give a stable dimer). On the other hand, di- and triphosphate form stable inclusion complexes. Triphosphate, as judged from available crystal structures of $H_3P_3O_{10}^{3-}$, shows an O---O distance as large as 7 Å, which may fulfil the cryptate's geometrical requirements. The success of diphosphate in removing the indicator from the cryptate may be surprising, considering the 'short' bite distance, never larger than 5 Å. However, in view of their lower bulkiness, di- and triphosphate may be really included inside the cryptate cavity, and rules of O---O and Cu^{II}---Cu^{II} matching should be completely different from those governing the cryptate–NMP interactions discussed above.

This qualitative study was extended to adenine polyphosphates ADP and ATP. Corresponding spectrofluorimetric titration profiles are shown in Figure 13.8, including those for AMP taken under the same conditions.

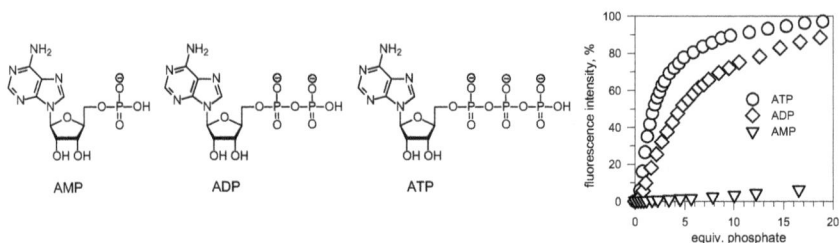

Figure 13.8. Spectrofluorimetric titration profiles obtained for adenine phosphates under the same experimental conditions described in the caption to Figure 13.7. The fluorescence intensity is normalised with respect to that of the uncomplexed indicator ($I/I_0 \times 100$).[103]

The fluorimetric response is quite similar to that given by inorganic phosphates illustrated in Figure 13.7. It has been already mentioned that AMP is forced to use the poorly donor aniline nitrogen atom to bind one metal centre of the cryptate, which explains the poor capability to displace fluorescein. ADP and ATP coordinate the two fixed Cu^{II} ions of the receptor by using two oxygens of the phosphate moiety. In the present case, the advantage of the triphosphate derivative over the diphosphate counterpart is more pronounced. This may be due to the fact that the cryptate adopts an open conformation similar to that illustrated in Figure 13.6, characterised by a larger bite, which favours the interaction with the longer triphosphate moiety.

14

Anion Inclusion by
Hexaprotonated Bistrens: Halides

It has been mentioned in Chapter 7 that protonated bistren can include anions.[73] In particular, the prototype of bistren cryptands, **22**, in aqueous solution adjusted at pH = 5 is present at ca. 80% as the hexaprotonated form LH_6^{6+} (the six secondary amine groups are protonated) and at ca. 20% as LH_5^{5+}. A solution of cryptand **22**, adjusted to pH 5, was titrated with a variety of anions and the progressive shifts of the ^{13}C resonances was monitored.[73] In any case, the formation of a complex of 1:1 stoichiometry was established. From the titration profiles (^{13}C NMR shift against anion equivalents), conditional association constants were determined: the highest value was observed in the case of N_3^- (log $K = 4.3 \pm 0.3$). The high stability was ascribed to the favourable fitting of the linear triatomic anion in the ellipsoidal cavity of the cryptand (see the crystal structure of the $[LH_6 \cdots N_3]^{5+}$ complex in Figure 7.3). Later, the complexation equilibrium (14.1)

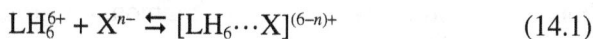

$$LH_6^{6+} + X^{n-} \leftrightarrows [LH_6 \cdots X]^{(6-n)+} \tag{14.1}$$

was investigated through pH-metric titrations. In particular, a solution of cryptand **22** containing excess acid was titrated with standard NaOH, first

in the absence and then in the presence of a chosen mononegative or dinegative anion and the varying equilibrium constants were determined
through a curve fitting procedure.[75] Thermodynamic constants for equilibrium (4) are shown in the bar plot in Figure 14.1.

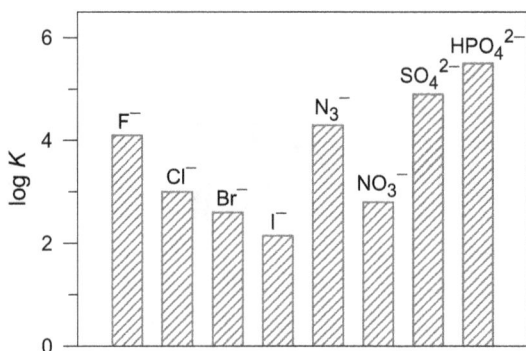

Figure 14.1. Log K values for the equilibrium $LH_6^{6+} + X^{n-} \leftrightarrows [LH_6 \cdots X]^{(6-n)+}$ in aqueous
0.1 M NaOTs, at 25°C (L = **22**).[75]

It is observed that, among halides, fluoride forms the most stable
complex and that stability decreases along the series $F^- \gg Cl^- > Br^- > I^-$.
Such a sequence reflects the decreasing tendencies of the anion to establish electrostatic interactions and to receive hydrogen bonds. In particular, it does not seem strictly related to geometrical factors and to the size/
shape matching between the anion and the receptor's cavity. In fact, the
crystal structure of the $[LH_6 \cdots F]^{5+}$ complex in Figure 14.2(a) demonstrates that the fluoride ion neither profits in full from the H-bond-
donating properties of the hexaprotonated cryptand, nor forces
receptor's framework to adapts its cavity to its geometrical requirements.[75] In particular, F^- chooses to receive hydrogen bonds from the
hydrogen atoms of four secondary ammonium group, according to a tet-
rahedrically distorted coordination geometry (Figure 14.2(b)). N---F
distances are those expected for the formation of rather strong hydrogen
bonds. The other two secondary ammonium groups lie at a definitely
non-bonding distances (5.7 Å).

The crystal structure of the corresponding chloride complex $[LH_6 \cdots$
$Cl]^{5+}$ in Figure 14.3(a) shows that the anion fully profits from the H-bond-
donating tendencies of the hexaprotonated receptor.[75]

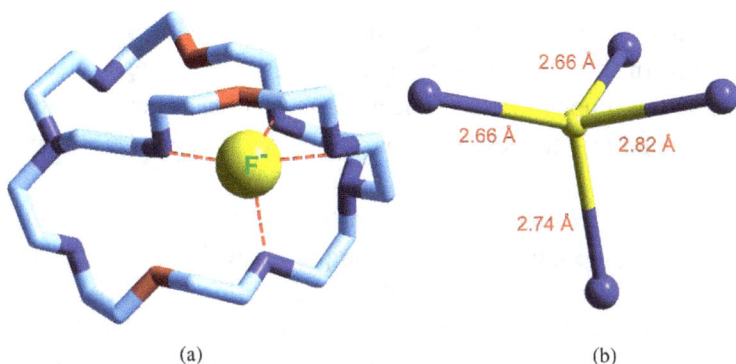

Figure 14.2. (a) The crystal structure of [LH$_6$···F](ClO$_4$)$_5$·0.5H$_2$O (perchlorate counteranions, solvating molecule and hydrogen atoms have been omitted for clarity; L = **22**)[75]; (b) distances between the fluoride ion (yellow sphere) and the four bound nitrogen atoms (F···H–N); nitrogen atoms are positioned at the corners of a distorted tetrahedron (hydrogens not represented).

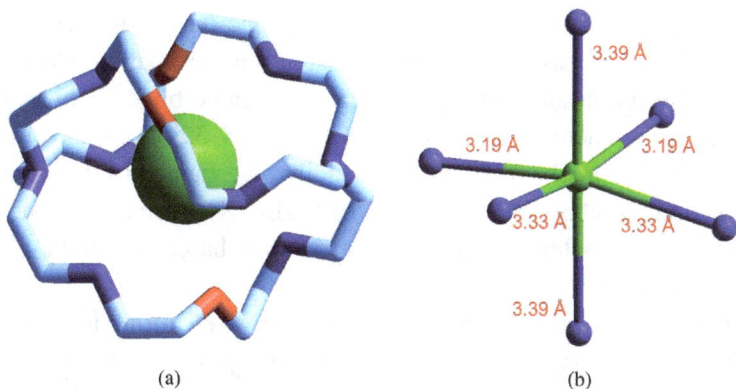

Figure 14.3. (a) The crystal structure of the [LH$_6$···Cl]Cl$_5$ (chloride counteranions have been omitted for clarity; L = **22**)[75]; (b) distances between the chloride ion (green sphere) and the six bound nitrogen atoms (Cl···H–N), positioned at the corners of a distorted octahedron (hydrogens not represented).

In particular, Cl$^-$ receives six hydrogen bonds from the six secondary ammonium groups of the hexaprotonated cryptand, according to a distorted octahedral coordination geometry. Thus, it may seem paradoxical that the chloride complex (6 H-bonds) is more than one order of magnitude less stable than that of fluoride (4 H-bonds). However, such a disadvantage is compensated by the higher intensity of the $F \cdots H-N$ interaction. In fact, F^-, in view of its highest electronegativity and its smallest radius among monoatomic anions, establishes the most energetic hydrogen bonding interactions with any $H-X$ fragment (X = non-metal atom).

The bar plot in Figure 14.1 indicates that the dinegative HPO_4^{2-} and SO_4^{2-} ions form the most stable 1:1 complexes, which suggests inclusion in the cryptand's cavity and establishing of multiple hydrogen bonding interactions. Lack of crystal structures prevents from a full understanding of the anion–receptor interactions. In any case, HPO_4^{2-} and SO_4^{2-} form more stable complexes than any other mononegative anion, either mono- (halides) or polyatomic (azide), which demonstrates the existence of selectivity mainly based on the electrical charge of the anion. The interaction of oxoanions with hexaprotonated cryptands will be discussed in detail in Chapter 15.

Also the study of anion binding by protonated bistren cryptands was extensively influenced by the invention of the one-pot synthesis involving the (3 + 2) Schiff base condensation of tren with aromatic dialdehydes, followed by the *in situ* hydrogenation of the imine bonds, which made available a complete class of octamines, e.g. **28–32**. Such a branch of anion coordination chemistry was especially investigated by Jane Nelson and coworkers, who synthesised tens of $[LH_6 \cdots X]^{5+}$ complexes, most of which were characterised by structural studies based on single-crystal X-ray diffraction.[107]

There exists a substantial difference in the anion inclusion by hexaprotonated cryptands, which depends upon the nature of the spacers linking the two bistren subunits, whether 'flexible' (e.g. $-CH_2CH_2OCH_2CH_2-$, **22**) or 'rigid' (those containing aromatic spacers, e.g. $-CH_2(C_6H_4)CH_2-$, *m-*, **31**, or *p-*, **32**). In fact, the 'flexible' hexammonium cryptand derived from **22** provides an elongated ellipsoidal cavity suitable for including the linear triatomic ion N_3^- (N_{tert}---N_{tert} distance: 8.84 Å, Figure 7.3), but it also includes monoatomic anions: for instance, it is able to contract its cavity in order to better accommodate chloride, establishing with the anion 6 H-bonds (N_{tert}---N_{tert} distance: 7.43 Å, Figure 13.4). The crystal structure of the neutral cryptand **22** is not available, but through molecular modelling a N_{tert}---N_{tert} distance of ~10 Å can be guessed, which suggests the occurrence of a drastic conformational rearrangement on protonation and complexation.

Cryptands containing an aromatic fragment in the spacer tell us a different story. Figure 14.4(a) shows the crystal structure of the neutral cryptand **29**.[108]

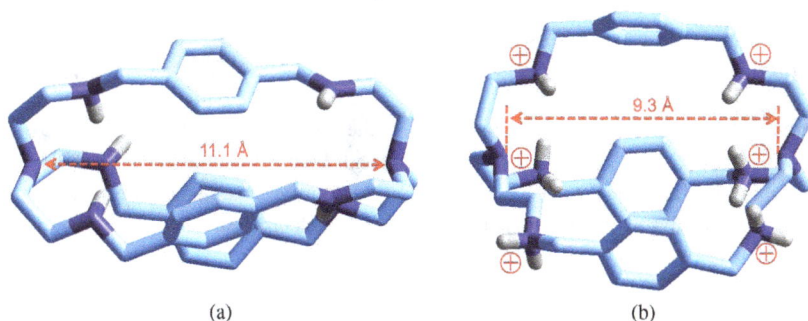

(a)   (b)

Figure 14.4. The crystal structures of: (a) $L \cdot 0.5H_2O$[108]; (b) $[LH_6](X)_2(HX)_2 \cdot H_2O$. $L = $ **29**, $X^{2-} = $ phthalate, $HX^- = $ hydrogenphthalate.[109] C–H hydrogens, solvating water and counteranions have been omitted for clarity.

The empty cryptand provides an ellipsoidal cavity, rather elongated, as indicated by the high N_{tert}---N_{tert} distance of 11.1 Å. On protonation of the six secondary amine groups, the length of the major axis diminishes to 9.3 Å.[109]

Bringing to pH 2 with HF 48% a methanolic solution of **29**, a salt with the formula $[LH_6]F_2 \cdot Cl_2 \cdot SiF_6$ was obtained as a white powder (L = **29**), which on recrystallisation from isopropanol and water yielded X-ray quality colourless plates. The crystal structure of the complex salt is reported in Figure 14.5(a).[110] The hexaprotonated cryptand encapsulates two F⁻, bridged by a water molecule. Each F⁻ receives four H-bonds (see Figure 14.5(b)): three from the N–H fragments of the secondary ammonium groups of one tren subunit and one from the bridging water molecule (N–H···F distances ranging from 1.69 to 1.83 Å; O–H···F distances 1.87 and 1.89 Å). The N_{tert}---N_{tert} distance (10.7 Å) is intermediate between that of the neutral cryptand (longer) and that of the hexaprotonated form (shorter). Noticeably, the SiF_6^{2-} counterions have been generated *in situ* through the attack of the glass container by hydrofluoric acid.

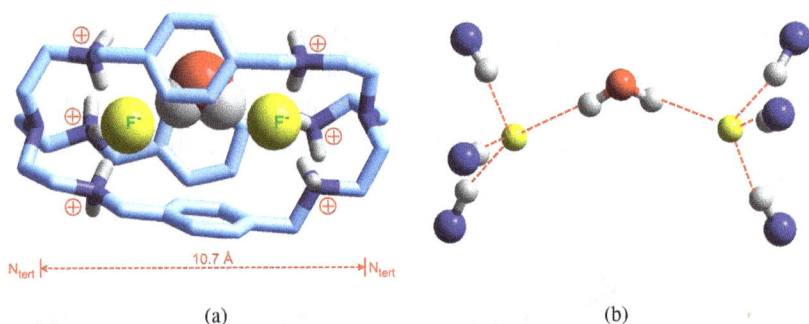

(a)   (b)

Figure 14.5. (a) The crystal structure of the complex salt [LH$_6$···F···H$_2$O···F] Cl$_2$(SiF$_6$)$_2$·12H$_2$O (L = **29**)[110]; (b) network of H-bonds holding together the F$^-$···HOH···F$^-$ moiety inside the hexaprotonated cryptand. Each fluoride experiences a roughly tetra-hedral coordination from three N–H and one O–H fragments.

It remains now to account for the different behaviour of the hexapro-tonated derivatives of **22** and of **29** when interacting with fluoride. The hexaprotonated cryptand **22**, probably due to the flexibility of its –CH$_2$CH$_2$OCH$_2$CH$_2$– spacers, contracts its cavity to accommodate fluo-ride and establishes four H-bonds, by using three N–H fragments from one tren subunit and one N–H fragment from the other. Such a rearrange-ment would involve a serious energy cost to the receptor containing the more rigid –CH$_2$(C$_6$H$_4$)CH$_2$– spacer, which therefore prefers to maintain its original conformation and to include two fluoride ions along with a bridging water molecule. Such a water molecule (i) shields the inter-ani-onic electrostatic repulsion and (ii) allows each F$^-$ to receive its preferred number of H-bonds (four, at the corners of a tetrahedron).

Something similar takes place in the interaction with chloride. Figure 14.6(a) shows the crystal structure of the complex salt [LH$_6$···Cl···H$_2$O]Cl$_5$·4H$_2$O·MeOH, which was obtained on addition of concentrated HCl to the methanolic solution of cryptand **29**.[111]

A chloride ion is well incorporated into the receptor's cavity: it receives three H-bonds from the three ammonium groups of one tren subunit and one H-bond from an included water molecule, whose oxygen atom receives in turn two H-bonds from two ammonium groups of the other tren subunit (see Figure 14.6(b)). The N$_{tert}$---N$_{tert}$ distance (10.1 Å)

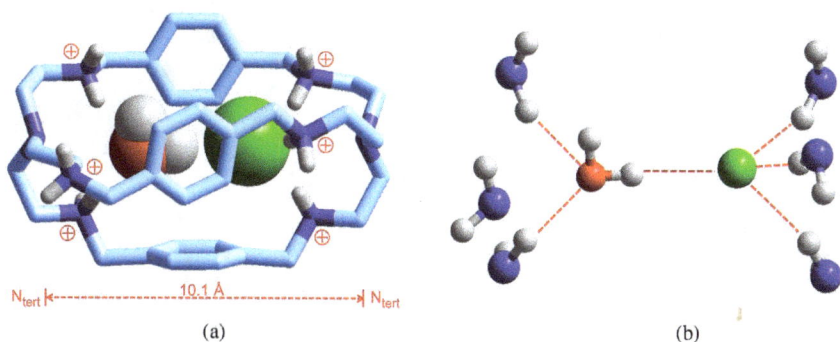

Figure 14.6. (a) The crystal structure of the complex salt [LH$_6$···Cl···H$_2$O] Cl$_5$·4H$_2$O·MeOH (L = **32**)[111]; (b) network of H-bonds holding together the Cl⁻···HOH group inside the hexaprotonated cryptand. The chloride anion experiences a roughly tetrahedral coordination.

attests for the elongated nature of the cavity, closer to the empty hexaprotonated receptor. Thus, also in the present case, the receptor chooses not to modify its conformation and to host the anion in a single tren subunit.

Bromide is similar to chloride in that it forms an [LH$_6$···Br···H$_2$O]$^{5+}$ complex, in which bromide receives three H-bonds from one triprotonated tren subunit and one from the included water molecule.

15

Anion Inclusion by Hexaprotonated Bistrens: Oxoanions

All the cryptands of the class **28–32**, in their hexaprotonated form, offer a comfortable shelter to tetrahedral oxoanions. Hydrogen bonding complexes with ClO_4^-, SO_4^{2-}, SeO_4^{2-}, CrO_4^{2-}, $S_2O_3^{2-}$ with most of the above-mentioned receptors have been isolated in the crystalline form and structurally characterised. Oxygen atoms of the oxoanions establish with the N–H fragments of the protonated secondary ammonium groups hydrogen bonding interactions of variable strength, characterised by different $O \cdots H$ length. At this stage, it is useful to consider the Jeffrey's classification of hydrogen bonds summarised in Table 15.1.[112]

H-bonds present in $[LH_6 \cdots XO_4]^{5+}$ inclusion complexes fall in the classes 'moderate' and 'weak'. In the crystal, water molecules are typically present. They are never included in the cavity and often, from the outside, bridge an N–H fragment and anion oxygen atom. Thus, they effectively contribute to the stability of the crystal, but cannot provide a realistic view of the hydration of the complex in solution. Therefore, water molecules establishing hydrogen bonding interactions with the complex

Table 15.1. Strong, moderate and weak hydrogen bonds following the classification of Jeffrey.[112]

	Strong	Moderate	Weak
Interaction type	Strongly covalent	Mostly electrostatic	Electrostat./dispers.
Bond length H$\cdots$A [Å]	1.2–1.5	1.5–2.2	>2.2
Distance X$\cdots$A [Å]	2.2–2.5	2.5–3.2	>3.2
Directionality	Strong	Moderate	Weak
Bond angles X–H$\cdots$A [°]	170–180	>130	>90
Bond energy [kcal mol^{-1}]	15–40	4–15	<4

Note: The interaction involves an X–H donor and an A acceptor. In the cases discussed in this chapter, X is the nitrogen atom of a secondary ammonium group of the hexaprotonated cryptand and A is the oxygen atom of an oxoanion.

will neither be shown in the structure reported in the figures below nor discussed in the text.

Figure 15.1(a) shows the crystal structure of the complex [LH$_6\cdots$ClO$_4$]$^{5+}$, L = **28**.[113]

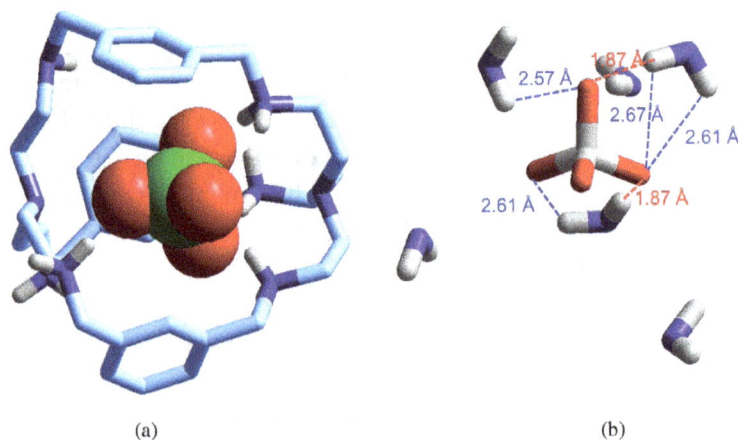

(a)   (b)

Figure 15.1. (a) The crystal structure of the complex salt [LH$_6\cdots$ClO$_4$](ClO$_4$)$_5\cdot$3H$_2$O, L = **28**[113]; perchlorate counteranions (not included), solvating water molecules, C–H hydrogen atoms have been omitted for clarity; the N$_{tert}$---N$_{tert}$ distance is 7.71 Å; (b) the hydrogen bonding network of the included perchlorate: red dashed lines, moderate hydrogen bonds; blue dashed lines, weak hydrogen bonds. The perchlorate anion receives (2 + 4) H-bonds.

The perchlorate anion is well incorporated in the cavity of the hexaprotonated cryptand. However, not all oxygen atoms of the anion receive H-bonds from N–H fragments (see Figure 15.1(b)). Incomplete hydrogen bonding coordination is a general feature of oxoanion cryptate complexes.

Figure 15.2(a) shows the structure of the $[LH_6 \cdots SeO_4]^{4+}$ complex (L = **28**).[114]

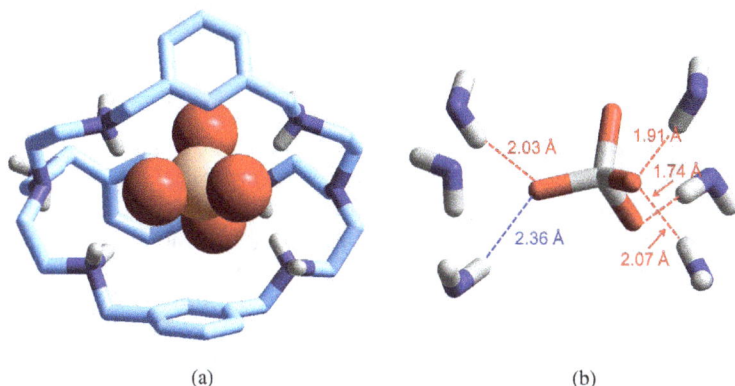

(a)   (b)

Figure 15.2. (a) The crystal structure of the complex salt $[LH_6 \cdots SeO_4](ClO_4)_4 \cdot H_2O$ (L = **28**)[114]; perchlorate counteranions, solvating water molecule, C–H hydrogen atoms have been omitted for clarity; the N_{tert}---N_{tert} distance is 9.21 Å; (b) the hydrogen bonding network of the included selenate: red dashed lines, moderate hydrogen bonds; blue dashed lines, weak hydrogen bonds. The selenate ion receives (4 + 1) H-bonds from the N–H fragments of five ammonium groups of the receptor.

In this case, the SeO_4^{2-} anion receives from the hexaprotonated cryptand five H-bonds, four moderate and one weak. Notice that three moderate H-bonds are donated by one tren subunit, one from the other (see Figure 15.2(b)).

A particular arrangement is observed for the nitrate complex, whose structure is shown in Figure 15.3(a).[115] The hexaprotonated form of cryptand **28** accommodates into its cavity *two* nitrate ions, parallel and eclipsed: each anion receives three moderate H-bonds from the three secondary ammonium groups of one tren subunit of the hexaprotonated cryptand, as shown in Figure 15.3(b). The formation of a network of six H-bonds may compensate the electrostatic repulsion between the two anions (the distance between the two nitrate nitrogens is 3.34 Å).

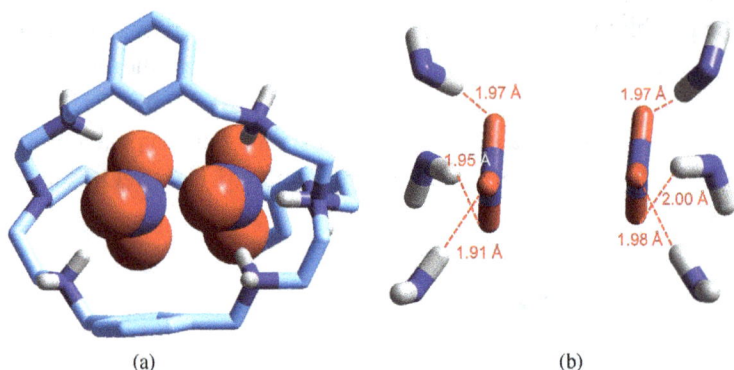

(a)   (b)

Figure 15.3. (a) The crystal structure of the complex salt $[LH_6 \cdots (NO_3)_2](NO_3)_4 \cdot 2H_2O$ (L = **28**)[115]; not included nitrate anions, solvating water molecule, C–H hydrogen atoms have been omitted for clarity; the N_{tert}---N_{tert} distance is 9.48 Å; (b) the hydrogen bonding network of the two incorporated nitrate anions: each nitrate receives 3 moderate H-bonds from the 3 secondary ammonium groups of one tren subunit of the hexaprotonated form of cryptand **28**.

The length of the major axis of the ellipsoid in oxoanion complexes varies according to the size of the incorporated anion. For the smaller perchlorate (average Cl–O bond distance 1.47 Å), N_{tert}---N_{tert} is 7.21 Å, distinctly lower than for bigger selenate (average Se–O bond distance 1.64 Å), in which N_{tert}---N_{tert} is 9.21 Å. Inclusion of the two nitrate ions does not require any special expansion of the cryptate cavity: N_{tert}---N_{tert} = 9.48 Å.

As far as the stability in solution of the hexaprotonated bistren/oxoanion complexes is concerned, there exist rather dispersed data, obtained under different conditions, which do not allow a safe and concluding comparison. A homogeneous study has been performed on the interaction of protonated derivatives of cryptands **28**, **30** and **31**. The investigation was carried out through pH-titration experiments on solutions of the cryptand in the absence and in the presence of the envisaged anion, over the 2–12 pH interval. The grouped bar diagram in Figure 15.4 reports log K values for the equilibrium (15.1),[107] where L = **28**, **30** and **31** and Y^{2-} is SO_4^{2-}, SeO_4^{2-} and $S_2O_3^{2-}$:

$$LH_6^{6+} + Y^{2-} \leftrightarrows [LH_6 \cdots Y]^{4+} \qquad (15.1)$$

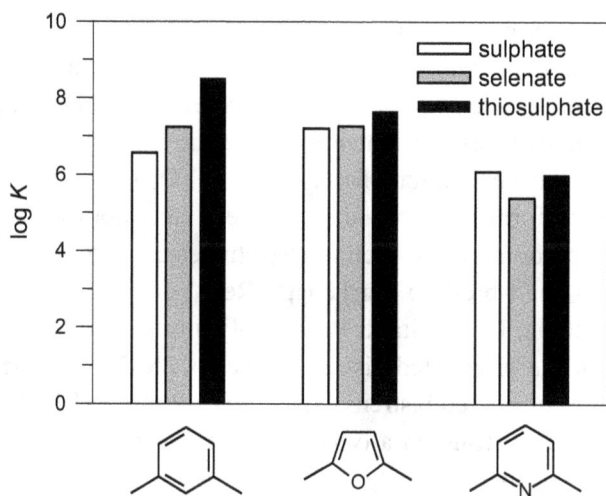

Figure 15.4. Log K values of the equilibrium $LH_6^{6+} + Y^{2-} \leftrightarrows [LH_6 \cdots Y]^{4+}$ in 0.1 M sodium tosylate at 25°C; L = **28**, **30** and **31** and $Y^{2-} = SO_4^{2-}$, SeO_4^{2-}, $S_2O_3^{2-}$. The spacers linking the two tren subunits of the cryptand are illustrated along the horizontal axis.[107]

The Log K values neither show a definite selectivity trend based on the nature of the anion, nor on the type of cryptand, a behaviour which may be related to the intrinsic weakness of the hydrogen bonding interaction. Several energy terms determine the stability of the oxoanion/receptor complex: one is the endergonic effect associated with the dehydration of the oxoanion. In this sense, the comparatively low hydration energy may account for the comparatively high stability of the thiosulphate complexes.

However, protonated bistren cryptands still represent the most powerful receptors for tetrahedral oxoanions in acidic aqueous solutions, a feature of special interest in the treatment of waste water from nuclear plants. In fact, among potential environmental pollutants present in the nuclear waste, there is the radioactive anion $^{99}TcO_4^-$. The long-lived isotope ^{99}Tc represents 6% of the total fission product yield. ^{99}Tc is a β-emitter (E_{max} = 293 keV, $t_{1/2}$ = 2.1 × 10^5 years), and, with the long-lived isotope ^{129}I (E_{max} = 194 keV, $t_{1/2}$ = 15.7 × 10^6 years), maintains the radioactivity of spent fuel for hundred thousands of years.[116] Due to the extremely high solubility in water of its sodium salt (11.3 M), $^{99}TcO_4^-$ migrates within the earth's crust and enters the food chain. Thus, $^{99}TcO_4^-$ uptake and removal is an important environmental task, the achievement of which is made difficult by the large size and the low charge the oxoanion.

The study of the solution chemistry of $^{99}TcO_4^-$ is not straightforward and must be performed in specialised nuclear laboratories, under strictly controlled conditions. Thus, preliminary studies are typically carried out on the $^{99}TcO_4^-$ half-brother: perrhenate, ReO_4^-. In a comparative study, Valeria Amendola, Università di Pavia, Italy, investigated the interaction of ReO_4^- with the protonated bistren cryptands 28–31 in an acidic solution.[117] All the considered bistrens exist at pH = 2 at 100% in the hexaprotonated form LH_6^{6+}. Thus, in a typical titration experiment, a solution of the envisaged cryptand was adjusted to pH 2 with triflic acid and titrated with $NaReO_4$. First, in a preliminary investigation, a ^1H NMR titration was carried out by adding $NaReO_4$ to a D_2O solution 7.7×10^{-3} M in 28, adjusted to pD 2 (see the titration profile in Figures 15.5(a) and 15.5(b), red triangles). Perrhenate addition induced a downfield shift of the signals corresponding to the $-CH_2-$ protons of the tren subunits of 28, a behaviour

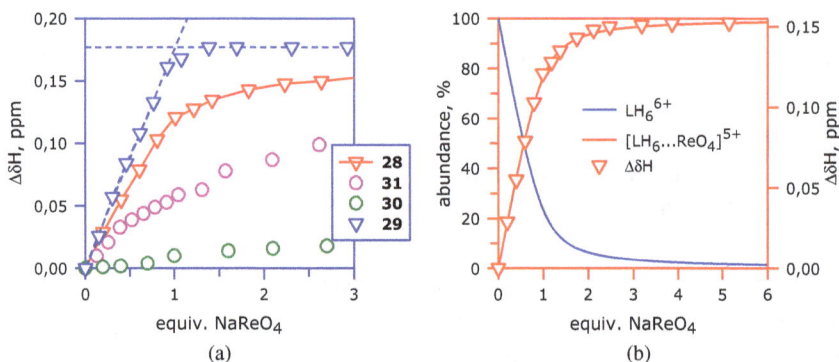

Figure 15.5 (a) Chemical shifts of the $-CH_2-$ protons of the tren subunits of cryptands 28–31 ($\Delta\delta H$) vs equivalent of $NaReO_4$ over the course of an ^1H NMR titration of a D_2O solution of the cryptand, adjusted to pD 2 with triflic acid; red solid line has been calculated for an association constant log $K = 3.4$; (b) symbols: chemical shifts measured during the ^1H NMR titration of a D_2O solution of cryptand 28, right vertical axis; lines: concentration (%) of the species at the equilibrium over the course of the titration, left vertical axis.[117]

consistent with the existence of significant electrostatic interactions between the methylene C–H protons and ReO_4^-. This unambiguously indicated the formation of a receptor/anion complex.

The smooth curvature of the titration profile allowed the safe determination of the binding constant. In particular, on fitting of ^1H NMR spectra by means of a dedicated nonlinear least-squares program, a log $K = 3.4 \pm 0.1$ (conditional constant, 0.1 M $NaCF_3SO_3$, pD = 2) was determined for the equilibrium (15.2):

$$LH_6^{6+} + ReO_4^- \leftrightarrows [LH_6 \cdots ReO_4]^{5+} \qquad (15.2)$$

However, in the case of cryptand **29** the curvature of the titration profile was too steep to allow a correct determination of the binding constant, whose value could be guessed to be $>10^5$ (see titration profile in Figure 15.5(a), blue triangles). On the other hand, for cryptand **31** no saturation profile was observed even on addition of a large excess of perrhenate, a feature that, at a cryptand concentration $\sim 10^{-2}$ M, corresponded to a log K value <2. In the case of cryptand **30**, poorly significant chemical shifts were observed, a behaviour which was ascribed to the presence of intramolecular hydrogen bonding interactions between the ammonium N–H fragments (donor) and the pyridine nitrogens of the spacers (acceptor). These intramolecular interactions made the inclusion of perrhenate into the cryptand's cavity especially difficult (endergonic).

Thus, the formation of perrhenate/cryptate complexes was investigated through potentiometric titrations in an aqueous solution made of 0.1 M in sodium triflate at 25°C. In particular, titrations were performed both in the absence and in the presence of 1 equivalent of the perrhenate and stepwise protonation constants log K_i were calculated. Difference of log K_6 values determined in the absence and in the presence of ReO_4^- gave log K for equilibrium (15.2). Corresponding values are reported in Table 15.2.

It is confirmed that receptor **29**, containing *p*-xylyl spacers, in its hexaprotonated form displays the highest affinity for ReO_4^-. In particular, the corresponding association constant is nearly two orders of magnitude higher than for the isomeric bistren **28**, which contains *m*-xylyl spacers.

Table 15.2. Log K values for the equilibrium $LH_6^{6+} + ReO_4^- \leftrightarrows [LH_6 \cdots ReO_4]^{5+}$, in an aqueous solution adjusted to pH 2, in 0.1 M $NaCF_3SO_3$. L = **28, 29, 30, 31**.

Method				
^{1}H NMR titration, log K	3.4 ± 0.1	>5	<2	<2
pH-titration, log K	3.5 ± 0.1	5.4 ± 0.1	<2	<2

Note: Receptors are indicated by the structural formulae of pertaining spacers.[117]

Complete thermodynamic information on the formation of ReO_4^-/cryptand complexes was obtained by means of isothermal titration calorimetry (ITC). Such a technique allows the simultaneous determination of (i) the association constant, log K, (ii) the corresponding enthalpy change, $\Delta H°$, and (iii) the variation of entropy, $\Delta S°$, calculated through the combination of $\Delta G°$ and $\Delta H°$, according to the Gibbs–Helmholtz equation (15.3):

$$\Delta G° = -RT \ln K = \Delta H° - T\Delta S° \tag{15.3}$$

The ITC technique, which requires minimum amounts of reactants, was devised for the study of binding of small molecules (e.g. a drug) to larger macromolecules (e.g. a protein),[118] but its use has been recently extended to the thermodynamic investigation of anion/receptor interactions.[119]

The instrument consists of two identical cells surrounded by an adiabatic jacket. Thermocouple circuits detect the temperature difference between the reference cell (filled with water) and the sample cell containing one of the reactants (e.g. the receptor). Each cell contains an electrical heater. On titration, known aliquots of the other reactant (e.g. the anion) are injected into the sample cell, generating a heat effect (either exo- or endothermic). If the reaction is endothermic, the system provides a power feedback to the heater in the sample cell in order to maintain an equal temperature in the two cells. If the reaction is exothermic, power feedback is provided to the reference cell. The observable quantity is the power needed to maintain the reference and the sample cell at an identical temperature. Power is recorded and plotted against time to give a series of spikes similar to those showed in the upper parts of Figure 15.6. Every spike (μcal s^{-1}) corresponds to one injection. Titration experiments are driven by a computer.

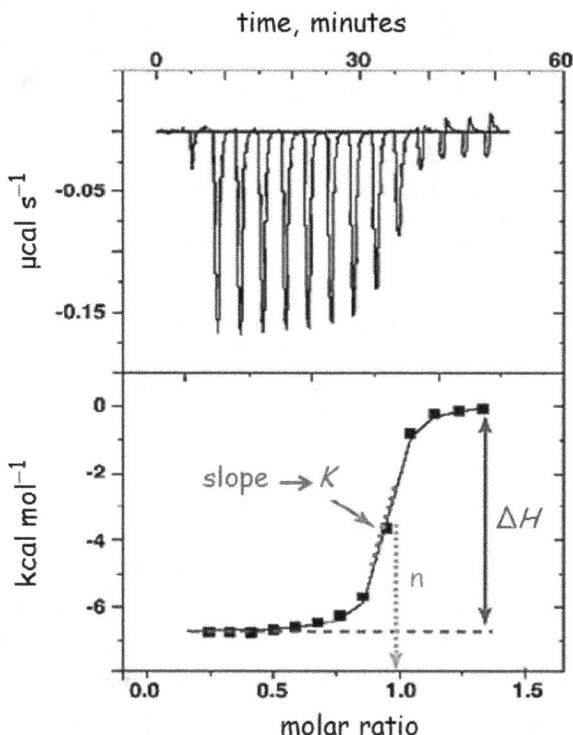

Figure 15.6. The diagram in the upper part of the figure shows the power spikes measured after each injection of the receptor (e.g. the cryptand) to the sample cell containing the substrate (e.g. the anion). In the lower part of the figure, the heat effect after each injection of perrhenate solution, obtained by integration of the corresponding power peak, is plotted vs the added equivalents From the profile, the number n, indicating the stoichiometry of the anion/receptor complex, and the complexation enthalpy are calculated.

A dedicated software integrates each power peak with respect to time to give the heat exchanged per injection (kcal mol^{-1}). A typical curve is shown in the lower parts of Figure 15.6. From the curve, the stoichiometry of the analyte/receptor complex (n) and the complexation enthalpy ($\Delta H°$) can be determined and also visually estimated. Computer elaboration of each curve provides complete thermodynamic information.

ITC experiments have been carried out to investigate the interaction of ReO_4^- with the hexaprotonated forms of bistren cryptands **28**, **29** and **31**, pH = 2, 0.1 M $NaCF_3SO_3$, 30°C. For L = **28** and **29**, the process is exothermic, for **31** athermic. The absence of any thermal effect following injection of ReO_4^- to an acidic solution of cryptand **31** confirmed the extremely poor stability of the corresponding complex (log K < 2) (Table 15.3).

Table 15.3. Thermodynamic quantities for the equilibrium: $LH_6^{6+} + ReO_4^- \leftrightarrows [LH_6 \cdots ReO_4]^{5+}$ (pH 2, 0.1 M $NaCF_3SO_3$, 30°C; L = **31**, **32**).

Method: ITC		
log K	3.29 ± 0.01	5.17 ± 0.01
$\Delta H°$, kcal mol^{-1}	-4.7 ± 0.1	-10.2 ± 0.1
$T\Delta S°$, kcal mol^{-1}	-0.1 ± 0.1	-3.1 ± 0.1

Note: Receptors are indicated by the structural formulae of pertaining spacers. Data determined through ITC.[117]

It is observed that the higher stability of the $[LH_6 \cdots ReO_4]^{5+}$ complex of **29** compared to **28** results from a substantially more exothermic enthalpy effect. Several terms concur to $\Delta H°$: (i) the formation of anion oxygens/N–H fragments hydrogen bonds (exothermic), (ii) the hydration of the $[LH_6 \cdots ReO_4]^{5+}$ complex (exothermic), (iii) the dehydration of the anion (endothermic), (iv) the dehydration of the hexaprotonated cryptand (endothermic) and (v) the conformational rearrangement of LH_6^{6+} following anion inclusion (endothermic). It is suggested that the hexaprotonated forms of cryptands **28** and **29** do not differ too much in hydration. Thus, selectivity cannot be ascribed to hydration terms. The role of hydrogen bond formation can be made clear by inspecting the crystal structures of the two complexes, shown in Figure 15.7.[120,117]

In each complex, the anion receives two H-bonds from the hexaprotonated receptors, a feature suggesting that it is not hydrogen bonding that determines the selectivity of the two cryptands. On the other hand, the conformational arrangement of the two complexes seems quite different. In the case of the hexaprotonated form of cryptand containing *p*-xylyl spacers, **29**, the anion is placed in the middle of the cavity (the distance

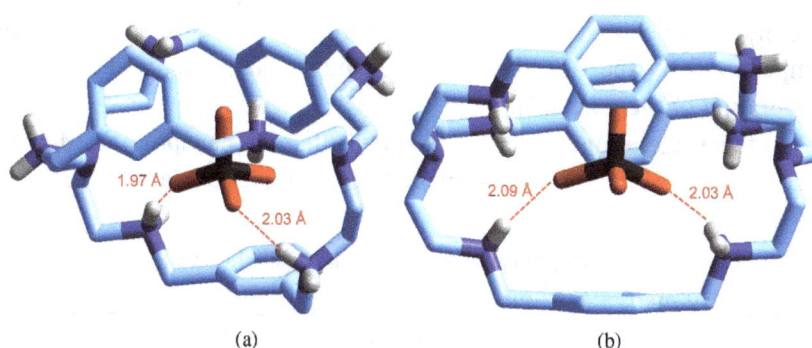

Figure 15.7. The crystal structures of: (a) $[LH_6 \cdots ReO_4](ReO_4)_5 \cdot 5H_2O$ (L = **28**)[120]; (b) $[LH_6 \cdots ReO_4](CF_3SO_3)_5 \cdot 8H_2O$ (L = **29**).[117] C–H hydrogen atoms, counteranions and solvational water molecules have been omitted for clarity.

between Re and the centroid of the receptor is 0.2 Å). Moreover, the structure of the complexed ligand is quite similar to that of the hexaprotonated bistren having a void cavity: in particular, the $N_{tert} \cdots N_{tert}$ distances are not too different, 9.32 Å (uncomplexed) and 9.85 Å (complexed). This indicates that, on anion inclusion, the receptor does not suffer any serious endothermic rearrangement. The situation is different in the case of the $[LH_6 \cdots ReO_4]^{5+}$ complex derived from the cryptand containing *m*-xylyl spacers, **28**. In fact, perrhenate does not occupy the centre of the receptor's cavity, the distance between Re and the centroid of the receptor being 0.6 Å. Moreover, on ReO_4^- inclusion, the cavity of the hexaprotonated receptor had to significantly contract: $N_{tert} \cdots N_{tert}$ distances: 9.64 Å (void), 8.69 Å (complexed). Thus, the nature of the spacer seems to play a significant role in determining the binding tendencies of the hexaprotonated bistren cryptand: in the presence of the more symmetric *p*-xylyl fragment, the receptor is prepared (*preorganised*, as Cram would say) to include ReO_4^- and to establish two H-bonds. The cryptand with the *m*-xylyl spacer, on the contrary, must undergo an endothermic conformational rearrangement, which is responsible for the distinct lowering of the complexation constant. In particular, the higher stability of the complex of cryptand **29** with respect to **28** is solely enthalpic in nature.

Entropy changes are more or less negative for both cryptate complexes (see Table 15.3). Disappearance of ReO_4^- from the solution accounts for a decrease of translational entropy, which is in part or totally compensated by the liberation of water molecules associated with the dehydration of the reactants.

At this stage, it must be ascertained whether the hexaprotonated form of cryptand **29** is the most convenient receptor for $^{99}TcO_4^-$ also.

Table 15.4. Thermodynamic quantities for the equilibrium: $LH_6^{6+} + X^- \leftrightarrows$ $[LH_6 \cdots X]^{5+}$ (pH 2, 0.1 M NaCF$_3$SO$_3$, 30°C; L = **29**).[121]

Anion, X^-	log K	$\Delta H°$, kcal mol^{-1}	$T\Delta S°$, kcal mol^{-1}	n
$^{99}TcO_4^-$	5.50 ± 0.01	−11.0 ± 0.1	3.3 ± 0.1	1.0
ReO_4^-	5.17 ± 0.01	−10.2 ± 0.1	3.1 ± 0.1	1.0
NO_3^-	3.41 ± 0.01	−3.4 ± 0.1	1.4 ± 0.1	0.9
Cl^-	2.25 ± 0.01	−1.4 ± 0.1	1.8 ± 0.1	0.9

Note: Data determined through ITC; n is the stoichiometric coefficient of the $[LH_6 \cdots X]^{5+}$ complex.

Thus, a thermodynamic investigation was carried out on the interaction of $^{99}TcO_4^-$ with LH_6^{6+} (L = **29**).[121] In particular, ITC titrations were performed by addition of a solution of the hexaprotonated cryptand to an $NH_4{}^{99}TcO_4$ solution adjusted to pH 2.0 with triflic acid and brought to 0.1 M ionic strength with NaCF$_3$SO$_3$, at 30°C. Pertinent thermodynamic quantities are reported in Table 15.4.

$^{99}TcO_4^-$ displays a behaviour similar to that of ReO_4^-: high stability of the $[LH_6 \cdots {}^{99}TcO_4]^{5+}$ complex, essentially due to a largely favourable enthalpy term. In particular, the pertechnetate complex is two times more stable than the perrhenate complex. Before exploring possible structural factors responsible for the moderately higher stability (0.3 log units), it is convenient to consider the hydration energies $\Delta H°_{hydr}$ of the two oxoanions: −60.0 kcal mol^{-1} for $^{99}TcO_4^-$ and −78.9 kcal mol^{-1} for ReO_4^-. Thus, the higher stability of the pertechnetate complex seems to derive essentially from the less endothermic dehydration process which precedes the anion encapsulation into the receptor's cavity.

Structural details of the $[LH_6 \cdots {}^{99}TcO_4]^{5+}$ complex,[121] shown in Figure 15.8, do not add any significant element for the understanding of its high thermodynamic stability. In particular, the presence of only one

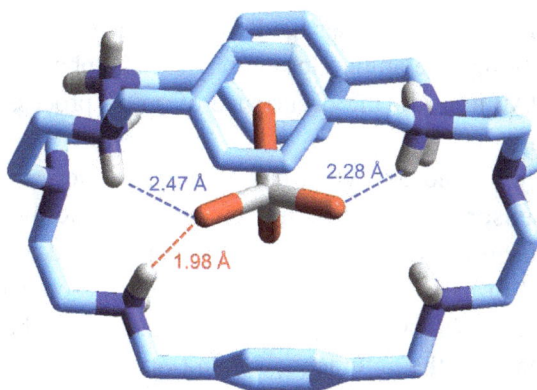

Figure 15.8. Crystal structure of the complex salt: [LH$_6$···^{99}TcO$_4$](TcO$_4$)(CF$_3$SO$_3$)$_4$·8H$_2$O (L = **29**).[121] C–H hydrogen atoms, counteranions and solvating water molecules have been omitted for clarity.

moderate hydrogen bond is observed (setting aside those established with outer water molecules, not represented in the figure). Based on these evidences, one should infer that a significant contribution, probably the most important, to the stability of the [LH$_6$···^{99}TcO$_4$]$^{5+}$ complex, as well to most complexes of oxoanions with hexaprotonated bistrens, comes from the electrostatic interaction between the anion and the six ammonium groups of the receptor. Thus, the more symmetrical the array of the positive charges, the greater the stabilisation of the anion/receptor complex.

Table 15.4 reports also thermodynamic quantities for the complexation by the same receptor of chloride and nitrate, possible competitors for the receptor in nuclear wastewater. The log K values show that [LH$_6$···Cl]$^{5+}$ and [LH$_6$···NO$_3$]$^{5+}$ are three and two orders of magnitude less stable than the pertechnetate complex, respectively, a circumstance which would allow effective separation and removal.

Recognition is strictly related to sensing. As shown in Chapter 11, a fluorescent sensor can be built, according to the FSR paradigm, by equipping a receptor with a fluorophore subunit: the essential prerequisite is that the envisaged substrate, when seized by the receptor, drastically modifies the emission properties of the fluorogenic subunit. In particular, it has been shown that the fluorescent emission of the dizinc(II) complex of bistren cryptand **34** (that containing two *m*-xylyl spacer and one

9,10-anthracenyl spacer) is quenched on addition of N_3^-. In particular, metal binding brings the anion close enough to the photoexcited anthracene subunit *An to allow the occurrence of an eT process and fluorescence quenching.

The same cryptand **34**, when hexaprotonated, acts as an ON/OFF fluorescent sensor of $^{99}TcO_4^-$ in an aqueous acidic solution.[122] Figure 15.9(a) shows the family of emission spectra taken in a solution of **34**, adjusted to pH = 2 with triflic acid, over the course of titration with a solution of $NaTcO_4$: on pertechnetate addition, the intensity of the emission band of the anthracene subunit progressively decreases.

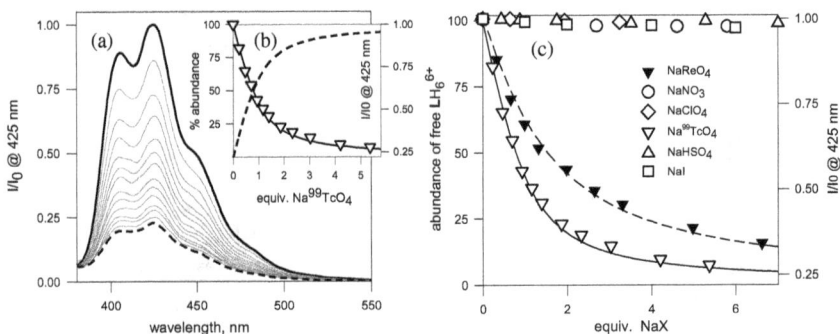

Figure 15.9. (a) Spectrofluorimetric titration of LH_6^{6+} (L = **34**; 10^{-5} M, 0.1 M $NaCF_3SO_3$, pH = 2.0) with $Na^{99}TcO_4$ at 25°C; (b) lines: concentration profiles of the species present at the equilibrium over the course of the titration (solid line, receptor; dashed line, complex); symbols: titration plot (normalised emission intensity, $I/I_0 \times 100$, at 425 nm vs equivalents of added anion): (c) symbols: titrations profiles for a variety of NaX salts, right vertical axis; lines: concentration profiles of LH_6^{6+} for perrhenate (solid line) and pertechnetate (dashed line), left vertical axis.[122]

Quenching implies anion inclusion into the receptor's cavity, whose occurrence has been demonstrated by the crystal structure of the complex salt $[LH_6 \cdots ^{99}TcO_4](CF_3SO_3)_5 \cdot 7H_2O$ (L = **34**), shown in Figure 15.10(a). Detailed inspection of the structural parameters shows that all the N–H$\cdots$O distances are ≥ 2.2 Å (one distance 2.17 Å), indicating the formation of only weak hydrogen bonding interactions. Thus, the high stability of the complex must be mainly ascribed to the establishment of especially strong electrostatic interactions. Indeed, the technetium atom is

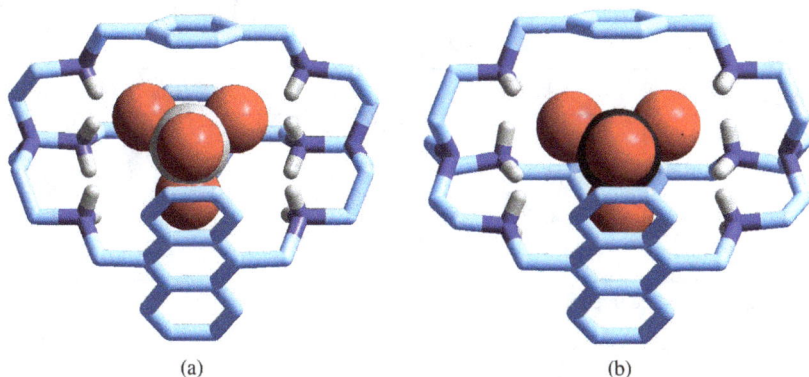

Figure 15.10. Crystal structures of the complexes: (a) $[LH_6\cdots^{99}TcO_4](CF_3SO_3)_5\cdot 7H_2O$ (L = **34**); (b) $[LH_6\cdots ReO_4](CF_3SO_3)_5\cdot 7H_2O$ (L = **34**). C–H hydrogen atoms, counteranions and solvational water molecules have been omitted for clarity.[122]

placed on the centroid of the hexaprotonated cryptand, at the same distance from all the nitrogen atoms of the six secondary ammonium groups (^{99}Tc---N = 4.31 ± 0.09 Å). Such a highly symmetrical arrangement favours the establishing of electrostatic interactions and is a probable consequence of the rigidity imparted to the receptor's framework by the 9,10-anthracenyl spacer.

The titration profile in Figure 15.9(b) shows that ReO_4^- also quenches the emission of the anthracene subunit present in the hexaprotonated cryptand. The $I/I_0 \times 100$, against anion equivalent profile for perrhenate is less steep than for pertechnetate, indicating the formation of a less stable complex. In any case, ReO_4^- is well incorporated in the hexaprotonated form of cryptand **34**, as shown by the crystal structure presented in Figure 15.10(b). Notice that the structure of the $[LH_6\cdots ReO_4]^{5+}$ complex is almost identical to that of the corresponding $[LH_6\cdots^{99}TcO_4]^{5+}$ complex: there exists one moderate H-bond interaction (N–H$\cdots$O distances 2.15 Å), but the symmetric placement of ReO_4^- inside the receptor's cavity (Re---N distances: 4.32 ± 0.07 Å) ensures the formation of rather strong electrostatic interactions.

The thermodynamic quantities associated with the formation equilibria of the $[LH_6\cdots X]^{5+}$ complexes (L = **34**, X = $^{99}TcO_4^-$, ReO_4^-, ClO_4^-, NO_3^-) were determined through ITC, and the corresponding values are reported in Table 15.5.

Table 15.5. Thermodynamic quantities for the equilibrium: $LH_6^{6+} + X^- \leftrightarrows [LH_6 \cdots X]^{5+}$ (pH 2, 0.1 M $NaCF_3SO_3$, 30°C; L = **34**).[122]

Anion	log K*	$\Delta H°$,* kcal mol^{-1}	$T\Delta S°$,* kcal mol^{-1}	log K (fluo)	log K (NMR)
$^{99}TcO_4^-$	5.49 ± 0.01	−13.4 ± 0.01	−5.8 ± 0.1	5.55 ± 0.01	—
ReO_4^-	5.20 ± 0.02	−9.15 ± 0.01	−1.9 ± 0.1	5.2 ± 0.1	—
ClO_4^-	3.61 ± 0.01	−6.50 ± 0.01	−1.5 ± 0.1	—	3.7 ± 0.1
NO_3^-	3.0 ± 0.1	−1.7 ± 0.1	—	—	3.2 ± 0.1

Note: * Data determined through ITC. All sodium salts.

Log K values are quite similar to those determined at 25°C through spectrofluorimetric or ^{1}H NMR techniques. It is observed, in particular, that the pertechnetate complex is two fold more stable than the perrhenate complex (Δlog $K = 0.29$), a feature common to derivatives of cryptand **32**. However, there exists a significant difference between the two cryptands: the formation of $^{99}TcO_4^-$ and ReO_4^- complexes for **34** is characterised by a distinctly more exothermic enthalpy term than for **29**. However, this advantage is exactly cancelled by a negative entropy contribution. Such a behaviour can be accounted for based on these structural features of the corresponding complexes: both $^{99}TcO_4^-$ and ReO_4^- complexes of **34** show a particularly well-ordered and symmetric structure, which allows the establishing of strong electrostatic interactions, which generates an especially exothermic effect. On the other hand, the formation of a well-ordered and rigid structure also has an unfavourable effect, that of producing a distinctly negative entropy term.

Going back to spectrofluorimetry, Figure 15.9(b) displays the titration profiles for a variety of inorganic anions: ClO_4^-, NO_3^-, HSO_4^-, I^-. The profiles are in every case flat, and anion addition barely modifies the emission of the anthracene subunit. Should one infer that these anions are not included in the receptor's cavity? Absolutely not. Anions enter the cavity and form relatively stable inclusion complexes, but they do not possess an appropriate mechanism to quench the nearby excited fluorophore. Thus, the question becomes: by which mechanism do $^{99}TcO_4^-$ and ReO_4^- quench the nearby excited anthracene subunit An*? There exist two possible mechanisms: electron transfer (eT) and energy transfer (ET). It seems convenient, at this stage, to outline how each mechanism operates.

Electron transfer (eT). The process is pictorially illustrated in Figure 15.11. The fluorogenic fragment anthracene (An) and an electron acceptor Ox (e.g. an oxoanion) have been brought to an interacting distance.

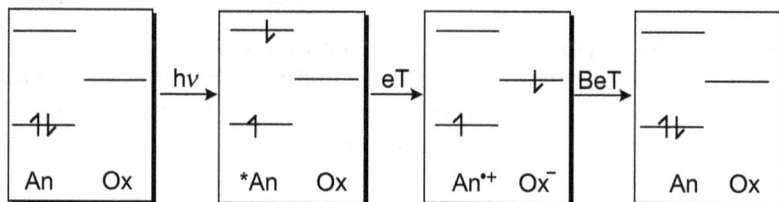

Figure 15.11. A molecular orbital scheme illustrating a photoinduced eT process, involving anthracene (An) and an electron acceptor (Ox). First, upon irradiation (*hv*), An is excited to *An. Then, the excited electron is transferred (eT) from *An to an empty orbital of Ox, of lower energy, giving rise to the radical cation $An^{•+}$ and to the reduced form Ox^-; at this point, inside the $\{An^{•+}, Ox^-\}$ion-pair, a back electron transfer (BeT) process takes place, restoring An and Ox. Through this mechanism, the photonic energy assumed by An is not restored by a fluorescent emission, but dispersed thermally.

First, upon excitation with a radiation of appropriate energy *hv*, one electron is raised from the HOMO (a π level) to the LUMO (a π^* level) of the fluorogenic subunit An, maintaining its spin multiplicity (singlet). Then, the excited electron is transferred from *An to an empty orbital of Ox (eT). Following this electron transfer process, an ion-pair forms, constituted by the radical cation $An^{•+}$ and by the reduced species Ox^-. Finally, a BeT process takes place inside the $\{An^{•+}, Ox^-\}$ion-pair, which restores An and Ox. Through this mechanism, the photonic energy assumed by the fluorophore is not restored in the form of a radiation, giving rise to the phenomenon of fluorescence, but is dissipated thermally.

Energy transfer (ET). This process is described in the molecular orbital scheme in Figure 15.12. According to this mechanism (proposed by D. L. Dexter in 1951),[123] the photonic energy assumed by the fluorophore An is transferred to a nearby molecular fragment X through a 'circular' movement of electrons, which leaves X in its excited state *X. The deactivation of *X can take place according to two distinct paths (i) radiative: the emission spectrum of X is observed, and not that of An: such a phenomenon is quite rare and precious — it is called *sensitised*

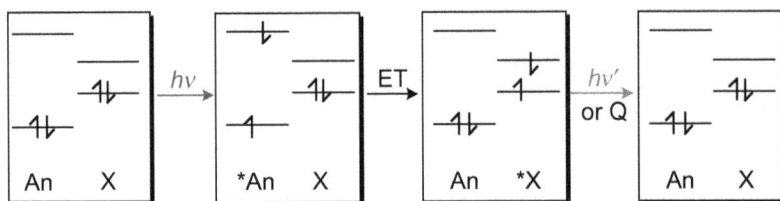

Figure 15.12. An ET process, according to the Dexter mechanism. Following the photonic excitation $\pi-\pi^*$ of one electron in An (hv), a circular transfer of electrons takes place involving both *An and X (electronic ET), which ends leaving X in its excited state *X. At this stage, *X can exhibit a radiative decay (hv') or, more frequently, a thermal decay (releasing a small amount of heat, Q).

luminescence; (ii) non-radiative: *X relaxes to its ground state thermally, i.e. by releasing a small amount of heat Q. Such a process, also called electronic ET or double electron exchange, requires that An and X are in close contact (in that differing from the Förster ET mechanism, which involves energy transfer through dipole–dipole coupling between species separated also by tens of Å).

How to distinguish between eT and ET mechanisms? There exists a rather simple empirical way: taking emission spectra at 77 K. Let us assume that we have an MeOH solution of the Fl–X conjugate and that X quenches the nearby fluorophore. The solution is frozen at the liquid nitrogen temperature and vitrifies (for this reason, MeOH has been used instead of water, which, under the same conditions, crystallises in the cuvette, thus preventing spectrofluorimetric studies): fluorescence is revived. Now, we can state that the quenching at room temperature in the liquid solution is due to an electron transfer process. Why?

Electrons move much faster than nuclei and, therefore, their motions can be separated (Born–Oppenheimer approximation). Thus, in a process involving the motion of both electrons and nuclei, it is assumed that nuclei first rearrange to the new coordinates, then electrons move as if nuclei were immobilised.

In Figure 15.13(a), the eT process occurring for an Fl–X conjugate system in a liquid solution at room temperature is pictorially illustrated. An eT process from *Fl to X generates the {Fl$^{•+}$–X$^-$}ion-pair. According to the separation of the motion of electrons and the motion of nuclei, occurrence of the process requires that solvent molecules first rearrange

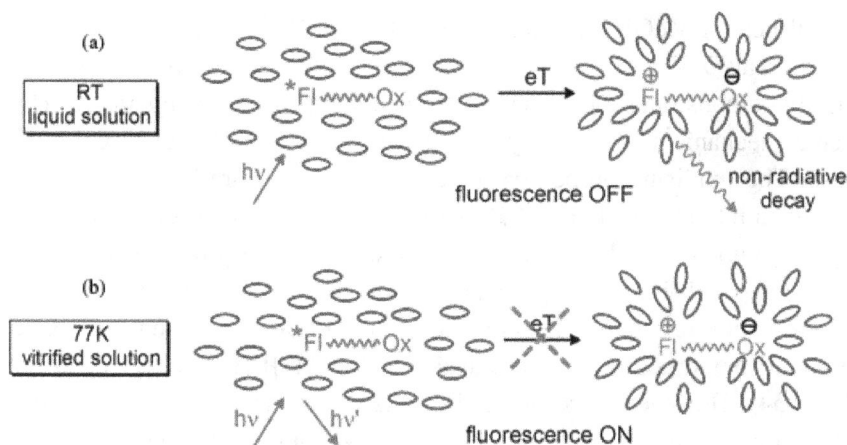

Figure 15.13. (a) In a liquid solution at room temperature the excited fluorogenic fragment *Fl of an Fl–Ox conjugate is quenched due to the occurrence of an eT to a close electron acceptor subunit Ox. The eT process generates charge separation and a rearrangement of the solvent molecules around the {Fl$^{\bullet+}$–Ox$^-$}ion-pair. Due to the Born–Oppenheimer approximation, first solvent molecules rearrange, then the electron is transferred; (b) the same process at 77 K: the solution is vitrified and solvent molecules cannot rearrange: the eT cannot take place and the typical fluorescent emission of Fl is released.

to assume the orientation demanded by the {Fl$^{\bullet+}$–Ox$^-$}ion-pair. When such an arrangement has been achieved, the eT takes place. At room temperature, nuclear motion (solvent molecules rearrangement) and electron motion (eT) occur according the required temporal sequence and fluorescence is quenched.

At 77 K things are different, as pictorially illustrated in Figure 15.13(b). Solvent molecules are immobilised and cannot rearrange to assume the orientation required by the {Fl$^{\bullet+}$–Ox$^-$}ion-pair. If nuclei do not move, the electron cannot move. The occurrence of the eT process is prevented and the normal fluorescence of Fl is released. Thus, fluorescence revival when the cuvette is thermostated at the liquid nitrogen temperature indicates that room temperature quenching is due to an eT mechanism. On the other hand, if quenching is maintained at 77 K, we are in the presence of an ET mechanism. In fact, the ET process involves the circular motion of electrons, with no charge separation and no solvent reorganisation.

When an MeOH solution of the $[LH_6 \cdots ReO_4]^{5+}$ complex (L = **34**) is vitrified at 77 K, anthracene fluorescence is fully restored, which unequivocally demonstrates the eT nature of quenching at room temperature. The same mechanism presumably operates in the $[LH_6 \cdots {}^{99}TcO_4]^{5+}$ complex also. The oxidising properties of ReO_4^- and ${}^{99}TcO_4^-$ are known. ClO_4^- is included into the cavity of LH_6^{6+}, but is not strong enough, as an oxidising agent, to uptake an electron from *An. However, it may be surprising that a strong oxidising agent like NO_3^- does not modify significantly the fluorescent emission of the anthracene subunit even if present in a large amount. No structural data are available for the $[LH_6 \cdots NO_3]^{5+}$ complex (L = **34**). However some useful insights can be taken from the crystal structure of the nitrate complex of the cryptand containing all *p*-xylyl spacers (L = **29**), shown in Figure 15.14(a).[124]

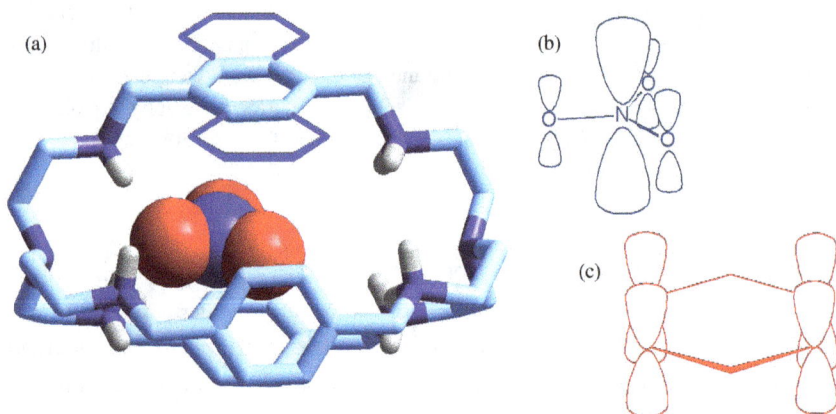

Figure 15.14. (a) The crystal structure of the complex salt $[LH_6 \cdots NO_3](CH_3C_6H_4SO_3)_5$ (L = **29**).[124] C–H hydrogen atoms and *p*-toluensulphonate counterions have been omitted for clarity; (b) the LUMO orbital of NO_3^- which receives an electron from the excited fluorophore; (c) the molecular orbital of a hypothetic phenyl fragment containing the excited electron. The eT process requires effective overlap of the two orbitals.

The nitrate ion is not placed in the centre of the cavity, but is positioned closer to one tren subunit. In particular, the distances of the NO_3^- nitrogen atom from the two bridgehead tertiary nitrogen atoms are 4.3 Å and 6.0 Å. Such an asymmetric arrangement allows the nitrate oxygen atom facing the closer tetrammonium subunit to receive two *moderate* hydrogen bonds from two secondary ammonium groups, which imparts a relatively high stability to the complex. However, such a structural arrangement is not favourable to the occurrence of an eT process from the *An subunit to the NO_3^- anion. In fact, the displacement of nitrate towards one tetrammonium subunit does not allow an effective overlap of the two involved molecular orbitals, sketched in Figures 15.14(b) and 15.14(c), which prevents eT and fluorescence quenching.

In conclusion, it can be stated that anion inclusion in hexaprotonated bistren cryptands is driven by both hydrogen bonding and electrostatic interactions. The formation of directional H-bonds is clearly observed for halides, which interact with a definite number of N–H fragments (4, 6), according to a given coordination geometry (tetrahedral, octahedral). Encapsulated oxoanions establish a lower number of directional hydrogen bonding interactions of moderate and weak intensity. This may depend upon the fact that the negative charge is spread over several oxygen atoms, whose basicity and affinity towards N–H protons is reduced. Thus, the electrostatic contribution, typically adirectional, is expected to play a major role. It derives that the design of selective polyammonium receptors for oxoanions, based on purely geometrical features, is especially problematic.

16

Recognition of Linear Dicarboxylates: The Hydrogen Bonding vs Metal–Ligand Interactions Contest

A reasonable question, at this stage, is: which, among the two classes of receptors derived from bistren, whether dicopper(II) cryptates or hexaprotonated cryptands, is the most effective and selective receptor for anions? Recognition of linear dicarboxylates is a convenient playfield where to test the potential of the two approaches.

In 1991, Lehn and coworkers considered that the hexaprotonated form of the previously discussed bistren cryptand (**26**)[76] possessed an ellipsoidal, oblong cavity suitable for the encapsulation of medium size dicarboxylates, whether aromatic or aliphatic. In particular, they crystallised the complex salt $[LH_6 \cdots 1,4\text{-}C_6H_4(COO)_2](C_6H_4(COO)_2)_2 \cdot 14H_2O$, in which a terephthalate ion is encapsulated within the hexaprotonated receptor LH_6^{6+}, (L = **26**). The structure of the inclusion complex is reported in Figure 16.1.[125]

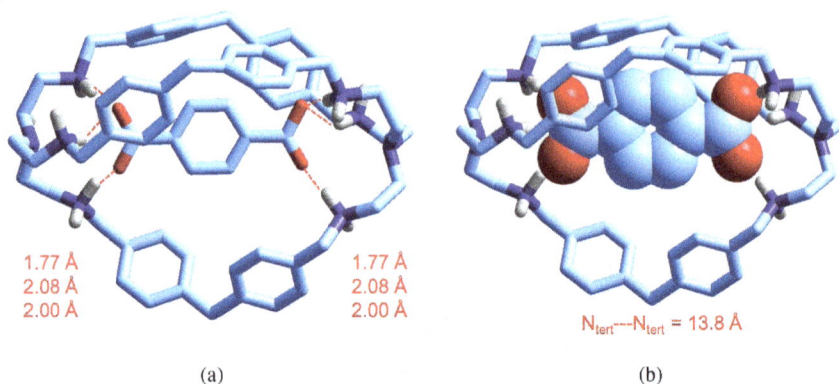

Figure 16.1. The structure of the terephthalate inclusion complex $[LH_6 \cdots C_6H_4(COO)_2]$ $(C_6H_4(COO)_2)_2 \cdot 14H_2O$ (L = **26**).[125] C–H hydrogen atoms, phthalate counterions and solvation water molecules have been omitted for clarity. Each carboxylate group establishes three moderate hydrogen bond with one triprotonated tren subunit, (bonds indicated by red dashed lines, N–H$\cdots$O distances reported in figure). The distance between the bridgeheads tertiary nitrogen atoms N_{tert}---N_{tert} is 13.8 Å.

The terephthalate dianion is included into the cavity of the receptor and establishes well-defined hydrogen bonding interactions with LH_6^{6+}. In particular, the oxygen atoms of each carboxylate group receives three H-bond of moderate strength from three N–H fragments of a triprotonated tren subunit, for a total of six. In a titration experiment, in a D_2O solution buffered at pD 6.0, at 20°C, the ^1H NMR signals of terephthalate were found to undergo marked upfield shifts on addition of LH_6^{6+}, indicating that complexation occurred. From data analysis, an association constant $K = 2.5 \times 10^4$ was calculated.[125] The study was then extended to a series of linear aliphatic dicarboxylates $^-OOC(CH_2)_nCOO^-$ (n = 2–8) and the corresponding formation constants of the 1:1 complexes were determined, whose values are reported in Table 16.1.

It should be noted that the association constants K for all the considered dicarboxylates are within only one order of magnitude. However, the subtle differences seem to disclose a meaningful behaviour, which is illustrated in Figure 16.2. In particular, the plot of K against the number of methylene groups in the carboxylic acid shows a well-defined peak selectivity behaviour, with the peak corresponding to n = 4 (adipate). It

Table 16.1. Log K values for the inclusion equilibria of linear aliphatic dicarboxylates of formula $^-OOC(CH_2)_nCOO^-$ (A^{2-}) by the hexaprotonated cryptand LH_6^{6+} in a D_2O solution at 20°C; buffered at pH = 6 with pyridine + CF_3COOD 10^{-2} M,[125] and by the $[Cu^{II}_2(L)]^{4+}$ complex in 50/50 (v/v) H_2O/EtOH at 25°C, pH = 7.2.[126] L = **26**.

$^-OOC(CH_2)_nCOO^-$	$LH_6^{6+} + A^{2-} \leftrightarrows$ $[LH_6 \cdots A]^{4+}$		$[Cu^{II}_2(L)]^{4+} A^{2-} \leftrightarrows [Cu^{II}_2(L)$ $(A)]^{2+}$	
n	K, M^{-1}	log K	K, M^{-1}	log K
2 (succinate)	1400	3.15	5.62×10^7	7.75
3 (glutarate)	2300	3.36	3.16×10^8	8.50
4 (adipate)	2600	3.41	1.02×10^{10}	10.01
5 (pimelate)	2100	3.32	2.19×10^7	7.34
6 (suberate)	1900	3.28	—	—
7 (azelate)	1400	3.15	—	—
8 (sebacate)	1500	3.18	—	—
Terephthalate (1,4-)	25000	4.40	6.17×10^9	9.79
Isophthalate (1,3-)	—	—	3.71×10^8	8.57
Phthalate (1,2-)	—	—	5.01×10^7	7.70

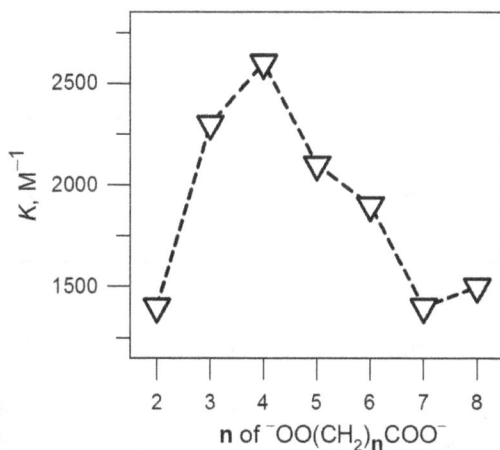

Figure 16.2. Plot of the constants K of the equilibrium: $LH_6^{6+} + A^{2-} \leftrightarrows [LH_6 \cdots A]^{4+}$ (L = **26**) vs the number of the methylene groups n of the α,ω-dicarboxylates $^-OOC(CH_2)_nCOO^-$ (A^{2-}), in a D_2O solution at 20°C; buffered at pH = 6 with pyridine + CF_3COOD, 10^{-2} M.[125]

was then suggested that the hexammonium receptor LH_6^{6+} (L = **26**) performs *linear recognition* of the substrate whose length matches that of the major axis of the ellipsoidal cavity provided by the hexaprotonated receptor.[125] Noticeably, the association constant for the adipate anion is one order of magnitude lower than that for terephthalate. In this regard, it must be considered that the flexible adipate anion on complexation is immobilised, so losing most of its degrees of freedom, a process involving a substantial decrease of entropy. Such an entropy loss is not observed on complexation of the rigid terephthalate, a circumstance favouring its inclusion. We are thus in the presence of a curious case of *preorganisation of the substrate* (in addition to that of the receptor).

More recently, Rita Delgado, Universidade Nova de Lisboa, and coworkers synthesised the dicopper(II) bistren cryptate $[Cu^{II}_2(L)]^{4+}$ (L = **26**) and investigated its interaction with medium size linear aliphatic dicarboxylates $^-OOC(CH_2)_nCOO^-$ (*n* = 2–4) and with the three positional isomers of phthalate.[126] The crystal structures of the corresponding inclusion complexes of terephthalate and of adipate are shown in Figure 16.3.

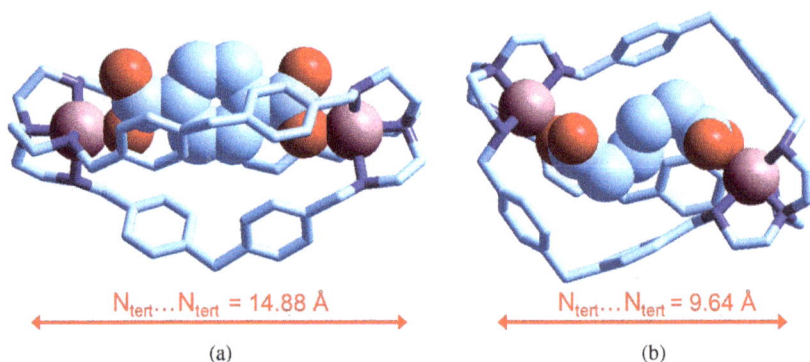

$N_{tert}\cdots N_{tert}$ = 14.88 Å $N_{tert}\cdots N_{tert}$ = 9.64 Å

(a)   (b)

Figure 16.3. The crystal and molecular structure of the salts: (a) $[Cu^{II}_2(L)(terephthalate)]$ $(ClO_4)_2 \cdot MeOH \cdot 4H_2O$ and (b) $[Cu^{II}_2(L)(adipate)](NO_3)_2 \cdot 10.8\ H_2O$ (L = **26**).[126] Hydrogen atoms, counterions and solvational molecules have been omitted for clarity. The length of the major axis of the oblong ellipsoidal cavity depends upon by the amplitude of the phenyl–CH_2–phenyl angle of each spacers: the larger the angle, the higher the distance.

It is observed that in the terephthalate inclusion complex (Figure 16.3(a)), the bistren framework is conveniently elongated in order to accommodate the rigid dianion. The distance between the bridgehead tertiary nitrogen atoms, N_{tert}---N_{tert}, 14.9 Å, is larger than that observed for the corresponding hexammonium terephthalate complex (13.4 Å). On the other hand, in the dicopper(II) adipate complex (Figure 16.3(b)), the N_{tert}---N_{tert} distance is much lower (9.6 Å), which indicates a great flexibility of the bistren framework. Such a flexible nature seems associated to the presence of a $-CH_2-$ group linking the two phenyl rings of each spacer. In particular, it is observed that N_{tert}---N_{tert} distance, which defines the length of the major axis of the receptor's ellipsoidal cavity, depends upon the amplitude of the phenyl–CH_2–phenyl angle of each spacer: the higher the amplitude of the angle, the higher the N_{tert}---N_{tert} distance. In particular, the terephthalate complex shows a more elongated ellipsoidal cavity (N_{tert}---N_{tert} distance: 14.9 Å) because the phenyl–CH_2–phenyl angles are larger (120.2°, 117.8°, 120.5°), compared to the less elongated adipate complex (N_{tert}---N_{tert} distance: 9.6 Å), exhibiting phenyl–CH_2–phenyl angles of distinctly smaller amplitude (112.2°, 112.4°, 111.3°).

The different elongation of the ellipsoidal cavity in the two cryptate complexes seems to derive from the geometrical requirements of the two dicarboxylates. If we consider the crystal structure of a purely ionic compound of terephthalate, the ammonium salt,[127] we observe that the distances of the *cis* oxygen atoms are 6.89 and 7.19 Å, values agreeing quite well with the distance of the coordinated oxygens in the cryptate: 6.95 Å. In the ammonium salt of adipate, the four O---O distances are 6.61, 6.60, 5.96 and 7.83 Å, to be compared with the distance between the coordinated oxygen in the cryptate: 6.50 Å. Thus, the dimetallic complex simply adjusts its organic framework to fulfil the geometrical requirements of each carboxylate.

The stability of the $[Cu^{II}_2(L)(A)]^{2+}$ complexes ($[L = $ **26**) with some aliphatic α,ω-dicarboxylates (A^{2-}) was investigated through potentiometric titrations in a 50/50 (v/v) H_2O/EtOH solution, at 25°C, pH = 7.2.[126] Table 16.1, right side, reports the values of the constants (K and log K) associated with the equilibria $[Cu^{II}_2(L)]^{4+} + A^{2-} \leftrightarrows [Cu^{II}_2(L)(A)]^{2+}$ ($[L = $ **26**) involving linear dicarboxylates $^-OOC–(CH_2)_n–COO^-$, $n = 2-5$.

The plot in Figure 16.4(a) (log K against n, triangles) shows a well-defined peak selectivity in favour of the adipate ion ($n = 4$), which forms an inclusion complex 2–3 orders of magnitude more stable than the

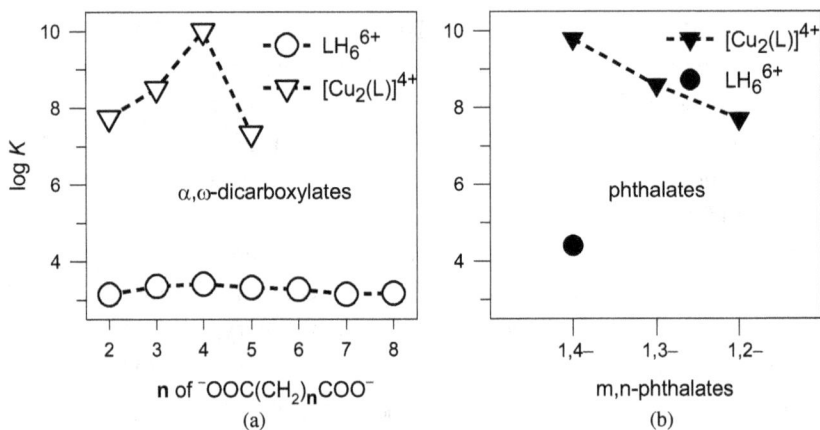

Figure 16.4. (a) Plot of log K values for the inclusion equilibria of dicarboxylates (A^{2-}) by the hexaprotonated cryptand LH_6^{6+} in a D_2O solution at 20°C (circles) buffered at pH = 6 with pyridine + CF_3COOD, 10^{-2} M,[125] and by the $[Cu^{II}_2(L)]^{4+}$ complex in 50/50 (v/v) $H_2O/EtOH$ at 25°C, pH = 7.2 (triangles),[126] L = **26**; linear aliphatic carboxylates of formula $^-OOC(CH_2)_nCOO^-$; horizontal axis, n; (b) positional isomers of m,n-phthalates; horizontal axis, m,n; triangles (inclusion by dicopper(II) cryptate) from Ref. 126; circle (inclusion by LH_6^{6+}) from Ref. 125.

corresponding anions of shorter (succinate, glutarate) and longer (pimelate) chain length. The plot illustrates well the capability of the dicopper(II) cryptate to recognise the length of the dicarboxylate. In the same Figure 16.4(a), values for the corresponding inclusion equilibria of the LH_6^{6+} receptor (L = **26**) are reported for comparative purposes.[125] This time, inclusion constants are expressed as logarithmic values. It is observed that, in the log scale, the selectivity of the LH_6^{6+} can hardly be detected. In any case, an absolute comparison of the two classes of data cannot be made, as equilibria were investigated in different media (pure water for LH_6^{6+} and 50/50 water/ethanol solution for $[Cu^{II}_2(L)]^{4+}$).

Equilibrium constants were also determined for the complexation by $[Cu^{II}_2(L)]^{4+}$ of the three positional isomers of m,n-phthalates (m,n = 1,4: terephthalate; 1,3: isophthalate; 1,2: phthalate), in a 50/50 water/ethanol solution.[126] Corresponding values are plotted in Figure 16.4(b). Log K values decrease along the sequence 1,4- > 1,3- > 1,2-, which parallels the distance of the potentially coordinating carboxylate oxygen atoms. Noticeably, the two champions of aliphatic and of aromatic dicarboxylates (adipate and terephthalate) have similar inclusion constants, in spite of the remarkably different conformation, contracted (adipate) and elongated (terephthalate), respectively (see Figure 16.3).

The results discussed above have demonstrated the superiority of dimetallic bistren cryptates in anion recognition with respect to hexaam-monium bistren cryptands. Such a higher capability of discrimination derives from the nature of the receptor–anion interaction. In particular, metal–ligand interactions are stronger than HB/electrostatic interactions and, in particular, they show a well-defined character of directionality. Such a feature is especially evident in transition metal ions, whose elec-tronic configuration (with partially filled d orbitals) may determine differ-ent coordination geometries. Directionality is much less pronounced in the case of hydrogen bonding and is totally absent for electrostatic interactions.

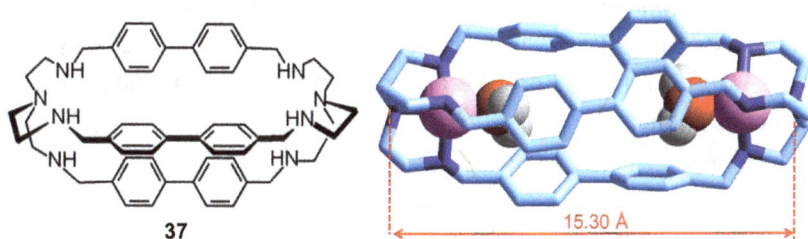

Figure 16.5. The crystal and molecular structure of the dimetallic complex salt $[Cu^{II}_2(L)(H_2O)_2](NO_3)_4$ (L = **37**).[128] Hydrogen atoms of the bistren ligand and counterions have been omitted for clarity. The distance between the oxygen atoms of the metal bound water molecules is 7.36 Å. The ideal dicarboxylate for the $[Cu^{II}_2(37)]^{4+}$ receptor should have its coordinating oxygens positioned at such a distance.

Bistren cryptand **37** is quite similar to **26**, as it provides an elongated cavity, but more rigid, due to the presence of 4,4′-ditolyl spacers. Very interestingly, the complex salt $[Cu^{II}_2(L)(H_2O)_2](NO_3)_4$ (L = **37**) was crystallised and its structure determined by X-ray diffraction studies.[128] Figure 16.5 shows the crystal structure of the 'void' cryptate, in which the fifth coordination site of each Cu^{II} centre is occupied by a water molecule. The structural arrangement offers a reliable representation of the cryptate receptor in water before anion inclusion. In view of the negligible steric hindrance of coordinate water molecules, the N_{tert}---N_{tert} distance (15.30 Å) is expected to correspond to that of the dimetallic receptor relaxed to its minimum energy conformation. On the other hand, the distance between the oxygen atoms of the metal-bound water molecules is 7.36 Å. We can therefore assume that the ideal dicarboxylate for the $[Cu^{II}_2(37)]^{4+}$ receptor should present its coordinating oxygens positioned at such a distance.

Bistren cryptand **37** and its dicopper(II) complex salts, including tetranitrate, have poor solubility in water ($<10^{-4}$ M). Thus, anion inclusion equilibria could not be safely investigated by potentiometric titration experiments, but had to be studied by the more sensitive spectrofluorimetric technique. In particular, the paradigm of the *fluorescent indicator displacement* was followed. Rhodamine, which possesses two carboxylate groups, and, when excited at 496 nm (isosbestic point), emits at 571 nm (orange fluorescence), was used as an indicator. In a typical experiment, the *chemosensing ensemble* solution was prepared with the dicopper(II) cryptate (2.5×10^{-6} M) and the indicator (2.5×10^{-7} M), whose fluorescence was completely quenched. Then, the chemosensing ensemble solution, adjusted to pH = 7, was titrated with a variety of dicarboxylates. Figure 16.6(a) shows the profiles obtained on titration with the positional isomers of phthalate.

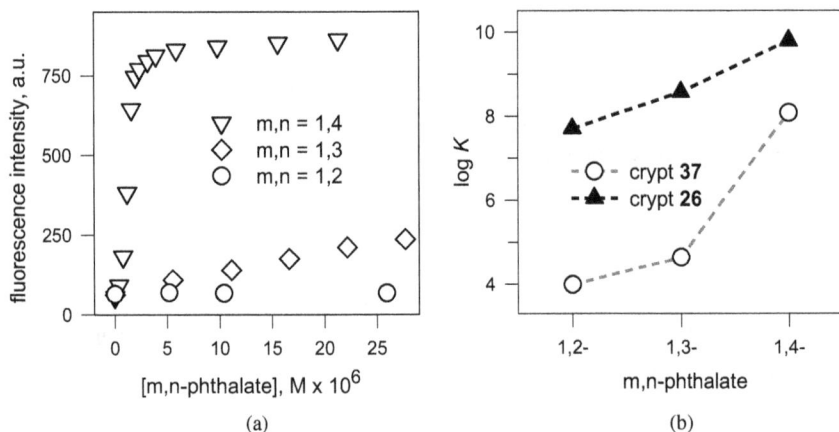

Figure 16.6. (a) Profiles obtained over the course of the titration of a solution containing the dicopper(II) cryptate $[Cu^{II}_2(L)]^{4+}$ (L = **37**, 2.5×10^{-6} M) and rhodamine (2.5×10^{-7} M) with an *m,n*-phthalate[128]; fluorescence emission at 571 nm on the vertical axis; only terephthalate (*m,n* = 1,4) fully displaces rhodamine and restores its orange fluorescence; (b) log K values for the equilibrium $[Cu^{II}_2(L)]^{4+} + A^{2-} \rightleftharpoons [Cu^{II}_2(L)(A)]^{2+}$: circles, L = **37**, water, pH = 7[128]; triangles, L = **26**, 50/50 water/ethanol.[126]

Titration profiles in Figure 16.6(a) show that only terephthalate is able to displace the indicator from the cryptate cavity, thus restoring the orange fluorescence of rhodamine. Isophthalate partially removes the indicator only when in large excess. Phthalate does not show any effect. This behaviour suggests that terephthalate should have two oxygen atoms at a distance close to 7.36 Å (O---O distance of the diaquo cryptate). Indeed, the distance of the oxygen atoms of ammonium terephtalate positioned *trans* each other is 7.39 Å.[127] Thus, the high stability of the complex depends upon the fact that two oxygens of terephthalate are able to replace the oxygens of the coordinated water molecules while barely modifying the structural arrangement of the cryptate.

In Figure 16.6(b), log K values of the of the inclusion equilibria of the three isomeric anions are reported and compared with those obtained for the dicopper(II) complex of cryptand **26**. The trend is the same, but an absolute comparison of the two classes of data cannot be done, due to the different media in which the systems were investigated. In particular, the higher stability observed for cryptates of **26** should mainly result from the lower degree of hydration of the anions in an H_2O/EtOH solution and to the corresponding lower endothermicity of the desolvation terms.

In any case, the $[Cu^{II}_2(L)]^{4+}$ receptor (L = **37**) shows a higher selectivity for terephthalate than the corresponding cryptate with L = **26**. This may be due to the higher rigidity of the 4,4′-ditolyl spacers, which make any conformational rearrangement especially endothermic. The $[Cu^{II}_2(L)]^{4+}$ receptor (L = **26**), containing the more flexible phenyl–CH_2–phenyl spacers, can accommodate isophthalate and phthalate at a lower energy cost.

The $[Cu^{II}_2(L)]^{4+}$ cryptate (L = **37**) also demonstrated a defined selectivity in the inclusion of linear aliphatic carboxylates. In particular, a neutral solution of $[Cu^{II}_2(L)]^{4+}$ (2.5×10^{-6} M) and rhodamine (2.5×10^{-7} M) was titrated with anions of formula $^-OOC–(CH_2)_n–COO^-$, n = 2–5. Titration profiles are shown in Figure 16.7(a).[128]

A well-defined selectivity in favour of dicarboxylates with n = 3 and 4 is observed. In fact, glutarate and adipate completely displace rhodamine from the dimetallic cryptate, whereas succinate and pimelate do not. Constants of the inclusion equilibria, calculated from spectrofluorimetric data, are shown in the diagram in Figure 16.7(b). The log K values for the dicarboxylate inclusion by LH_6^{6+} (L = **26**) are also reported.[125] Comparison is this time homogeneous in the sense that log K

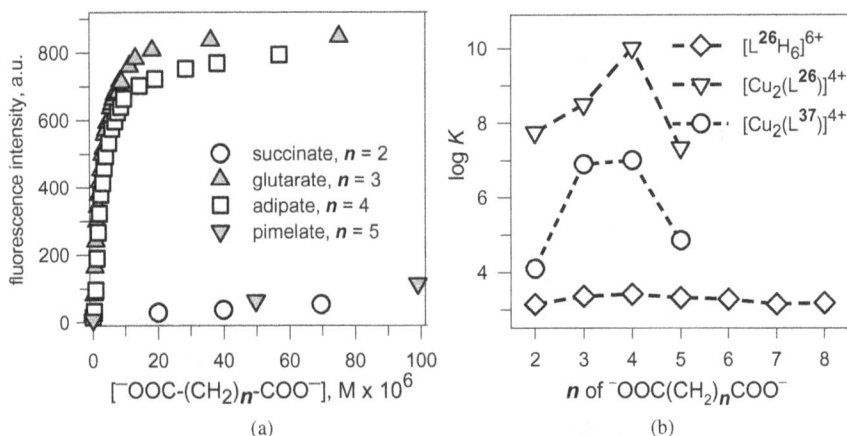

Figure 16.7. (a) Profiles obtained over the course of the titration of a solution containing the dicopper(II) cryptate $[Cu^{II}_2(L)]^{4+}$ (L = **37**, 2.5×10^{-6} M) and rhodamine (2.5×10^{-7} M) with linear dicarboxylates of formula $^-OOC-(CH_2)_n-COO^-$, n = 2–5[128]; fluorescence emission at 571 nm on the vertical axis, arbitrary units; both glutarate and adipate displace rhodamine and restore its orange fluorescence; (b) circles: log K values for the equilibrium $[Cu^{II}_2(L)]^{4+} + A^{2-} \rightleftharpoons [Cu^{II}_2(L)(A)]^{2+}$ (L = **37**) water, pH = 7[128]; triangles pertain to the equilibrium involving the $[Cu^{II}_2(L)]^{4+}$ cryptate (L = **26**, 50/50 water/ethanol)[126]; diamonds refer to the equilibrium : $LH_6^{6+} + A^{2-} \rightleftharpoons [LH_6\cdots A]^{4+}$ (L = **26**), in water at pH = 6.[125]

values were obtained in the same medium (pure water). It must be noticed that glutarate and adipate dimetallic complexes are 10,000 times more stable than the corresponding complexes of the hexaprotonated cryptand. This stability advantage almost disappears for anions in the presence of a shorter (succinate) or of a longer (pimelate) spacer. This definitively confirms the superiority of metal–ligand interactions with respect to HB/electrostatic interactions in anion recognition in water.

It is interesting to note that L-glutamate has (i) the same carbon framework as glutarate and (ii) the same oxygen–oxygen bite as glutarate.

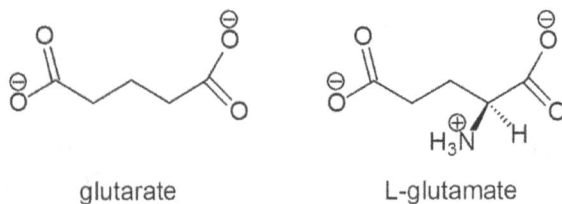

glutarate   L-glutamate

Thus, it is expected to give a similar fluorescent response to the dicopper(II) cryptate/rhodamine chemosensing ensemble. This issue is important from the perspective of cell physiology studies. In fact, L-glutamate is a major excitatory transmitter in the central nervous system,[129] and its selective determination in the presence of other neurotransmitters is strongly required by neurophysiologists. In particular, L-glutamate concentration must be determined in real time and real space, a feature provided by investigations with a fluorescence microscope.[130] In this regard, a chemosensing ensemble solution containing the dicopper(II) cryptate $[Cu^{II}_2(L)]^{4+}$ (L = **37**, 2.5 × 10^{-6} M) and rhodamine (2.5 × 10^{-7} M), adjusted to pH = 7 with HEPES, whose fluorescence was completely quenched, was titrated with a solution of L-glutamate.[131] The corresponding titration profile is displayed in Figure 16.8 (red symbols). Notice that the neurotransmitter concentration (horizontal axis) is expressed in log units.

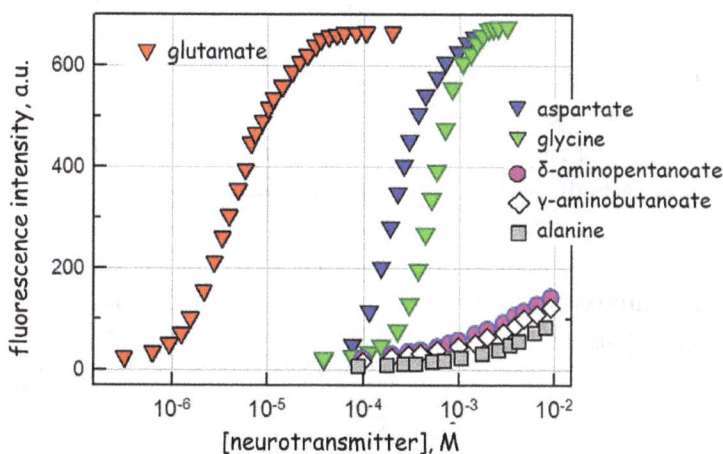

Figure 16.8. Profiles obtained over the course of the titration of a solution containing the dicopper(II) cryptate $[Cu^{II}_2(L)]^{4+}$ (L = **37**, 2.5 × 10^{-6} M) and rhodamine (2.5 × 10^{-7} M), adjusted to pH = 7 with HEPES buffer, with most important neurotransmitters; fluorescence emission at 571 nm on the vertical axis, arbitrary units.[131]

On processing fluorimetric data, a log *K* value of 6.9 ± 0.2 was determined, which is coincident with that of glutarate. This confirms the key role of the matching of Cu^{II}---Cu^{II} and O---O distances in the anion recognition by dicopper(II) cryptates.

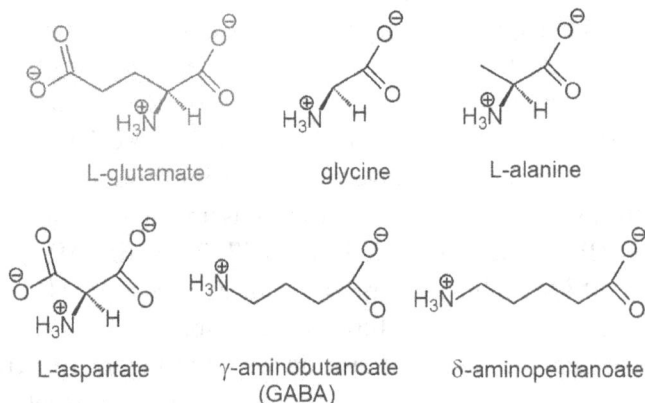

L-glutamate   glycine   L-alanine

L-aspartate   γ-aminobutanoate (GABA)   δ-aminopentanoate

Other relevant neurotransmitters are not expected to compete significantly for the dicopper(II) cryptate receptor. L-aspartate possesses two carboxylate groups, but they are separated by only one –CH_2– group. All the other neurotransmitters contains only one carboxylate group. Indeed, titration profiles in Figure 16.8 show that, at the lower concentration levels, the profile for L-glutamate is separated by 2 log units from that of L-aspartate, thus indicating high discrimination. The profile of glycine is surprisingly close to that of L-aspartate, which suggests that two molecules of the amino acid may interact with the two metal centres of the receptor. All the other monocarboxylate neurotransmitters barely interfere with L-glutamate fluorescence sensing.

17

Bistren Amides: Neutral Receptors that Recognise Anions only Through Hydrogen Bonding Interactions

The N–H fragment of the ammonium group can donate a hydrogen bond to an anion, but the N–H fragment of the parent primary or secondary amine cannot because it is not polarised. On the other hand, an N–H fragment of a primary or a secondary amide, irrespective of whether it is carboxamide or sulphonamide, is polarised enough to behave as an H-bond donor and to form a complex with anions. In general, such an interaction is not strong enough to compete with water. Thus, the anion amide complexes typically form in aprotic solvents, even if they are of pronounced polarity (e.g. DMSO, MeCN). Figure 17.1 illustrates how primary triamides of the branched tetramine tren can coordinate a given anion X^- by donating three convergent hydrogen bonds.

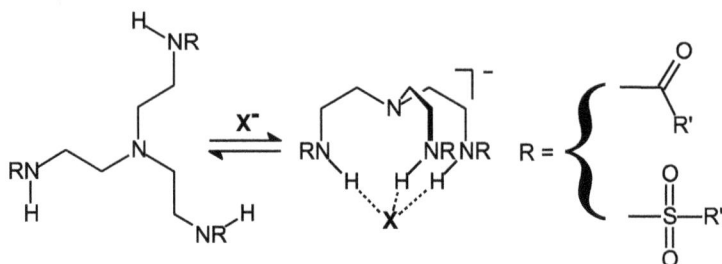

Figure 17.1. Amide derivatives of the branched tetramine tren. N–H fragments of the three secondary amide groups may behave as H-bond donors towards a given mono- or polyatomic anion X⁻. One could reasonably expect that the uncomplexed triamide adopts an open unwrapped conformation and, in the presence of X⁻, folds to assume a tripodal arrangement and to form the complex, thus losing entropy. Such a hypothesis will be proved wrong.

In 1993, David Reinhoudt, University of Twente, The Netherlands, prepared a series of amides of the tren family and investigated their binding tendencies towards anions in MeCN.[132]

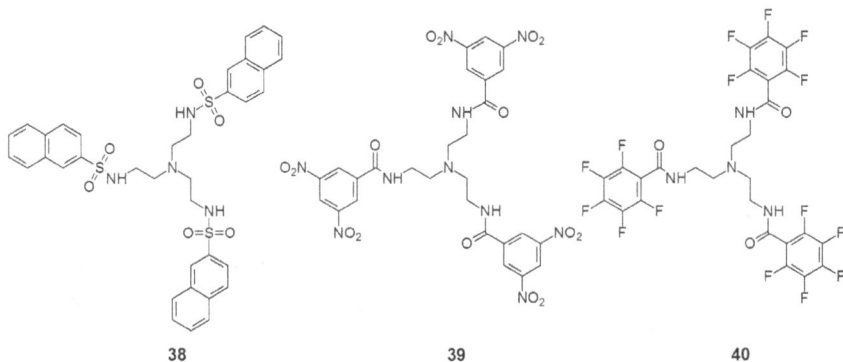

38   39   40

In particular, he observed that the tri-sulphonamide **38** forms an especially stable complex in MeCN ($\log K = 4.2$) with $H_2PO_4^-$, which is much more stable than the corresponding complexes of Cl⁻ and HSO_4^-. Since then, a variety of studies have been carried out to investigate the formation of anion complexes from amide derivatives of tren.

Figure 17.2(a) shows the crystal structure of the tren-based amide receptor **39**.[133] Quite unexpectedly, the uncomplexed receptor **39** shows a closed, tripodal structure, which may seem perfectly preorganised for the inclusion of a monoatomic anion. Indeed, such a favourable

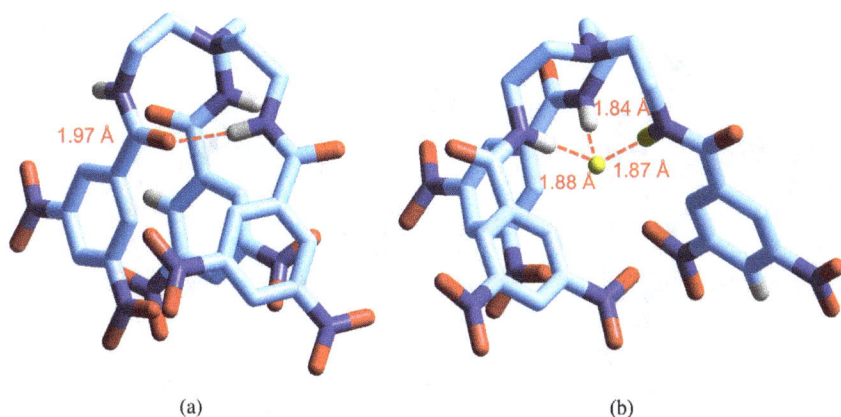

Figure 17.2. The crystal structures of (a) L·DMSO (L = **39**),[133] where the tripod is tightly closed due to the presence of an intramolecular H-bond, indicated by a red dashed line, and (b) [Bu₄N][L···F]·H₂O (L = **39**),[133] where fluoride receives three H-bonds from the hydrogen atoms of the three amide groups; fluoride complexation opens the tripodal cleft. C–H hydrogen atoms, solvating molecules and tetra-*n*-butylammonium cation have been omitted for clarity.

conformation results from the presence of an intramolecular hydrogen bond of moderate strength between an amide N–H fragment and the oxygen atom of an amide group belonging to another arm of the tren subunit. Thus, anion complexation must involve the endergonic breaking of this hydrogen bonding interaction, which may cancel, totally or partially, the advantage of a conformational preorganisation. Noticeably, all the reported crystal structures of carboxyamides and sulphonamides of tren show a closed conformation due to the presence of an intramolecular N–H···O=X hydrogen bonding interaction (X = C, S).

Receptor **39** forms a stable complex with fluoride, whose structure is shown in Figure 17.2(b).[133] In particular, F⁻ occupies the upper room of the tripodal cavity, where it receives three H-bonds from three properly oriented amide N–H fragments. On complexation, the tripod significantly opens its tight cleft, in order to accommodate fluoride.

The electron-withdrawing groups (–NO₂) in the phenyl rings of receptor **39** increase the acidity and enhance the H-bond donor tendencies of the amide N–H fragments (according to the accepted view of hydrogen bonding as a more or less advanced and 'frozen' proton transfer from the donor to the acceptor).[134] Thus, it is not surprising that the

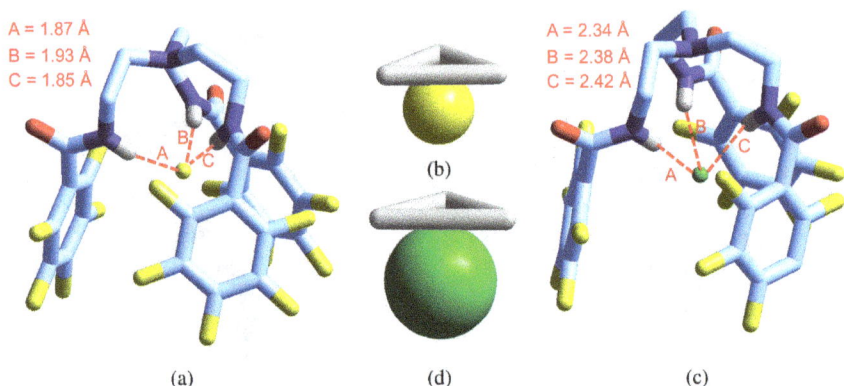

A = 1.87 Å
B = 1.93 Å
C = 1.85 Å

A = 2.34 Å
B = 2.38 Å
C = 2.42 Å

(a) (d) (c)

Figure 17.3. The crystal structures of the complex salts: (a) [Bu$_4$N][L$\cdots$F]·CHCl$_3$, and (c) [Bu$_4$N][L$\cdots$Cl]·CHCl$_3$, (L = **40**).[135] C–H hydrogen atoms, tetra-*n*-butylammonium and chloroform solvate have been omitted for clarity. Red dashed lines indicate the hydrogen bonding interactions of moderate strength involving N–H fragments and halide; (b) and (d) show triangles at whose vertices the amide hydrogen atoms interacting with the halide ion are positioned.

tren amide **40**, containing pentafluorobenzamide subunits, forms stable complexes with fluoride and chloride.[134] Their crystal structures are shown in Figure 17.3.[135]

The halide occupies the upper room of the tripodal cleft and establishes definite hydrogen bonding interactions with the amide hydrogen atoms, which are placed at the vertices of a nearly equilateral triangle (Figures 17.3(b) and 17.3(d)). Association constants of the order of 10^3–10^2 were determined through ^{1}H NMR titration experiments in CDCl$_3$, which displayed the typical sequence: F$^-$ > Cl$^-$ > Br$^-$ > I$^-$. Complexes like these do not form in water and other protic solvents because the rather weak receptor–substrate interaction energy cannot compensate the highly endergonic anion desolvation.

Well aware of the advantages of genuine preorganisation that do not arise from intramolecular interactions, Kristin Bowman-James, University of Kansas, Lawrence, synthesised a class of bistren cryptands containing secondary amide groups that are suitable for anion inclusion and formation of complexes held together by pure hydrogen bonding interactions.[136] Her first examples referred to molecules **41** and **42**.[137]

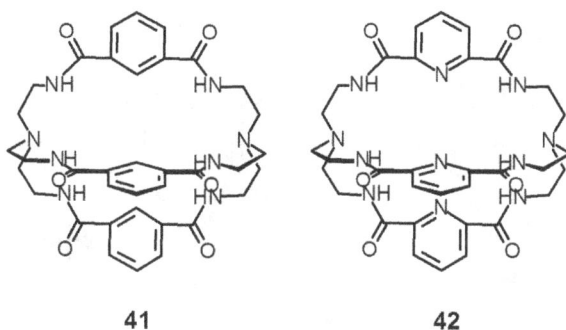

41   42

Polyamide cryptands and analogous neutral polycyclic systems have been proven as effective receptors with the capability to include a variety of anions, only in media unable to donate H-bonds. The bicyclic amide cryptand **42** was obtained in CH_2Cl_2 through the condensation of two equivalent of tren with three equivalent of 2,6-pyridinedicarbonyl dichloride in the presence of Et_3N as a base, as pictorially illustrated in Figure 17.4.

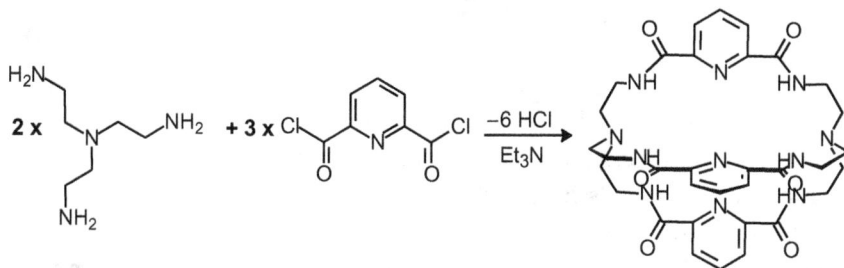

Figure 17.4. Synthesis of the bistren amide cryptand **42**. The irreversible nature of the primary amine–acyl chloride condensation limits the yield to 10%.[137]

42 was isolated in 10% yield after column chromatography (neutral Al_2O_3, 5% CH_3OH in CH_2Cl_2). The low yield was probably due to the irreversible nature of the primary amine acylation, which does not allow any trial and error mechanism, as observed in the synthesis of bistren amine cryptands through the Schiff base condensation.

Figure 17.5 shows the crystal structure of the complex salt $[Bu_4N]$ $[L\cdots F]\cdot MeCN\cdot CH_2Cl_2$ (L=**42**).[137]

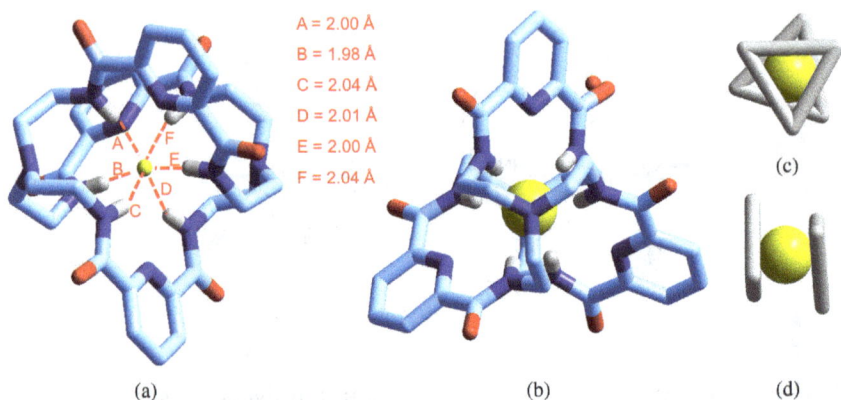

A = 2.00 Å
B = 1.98 Å
C = 2.04 Å
D = 2.01 Å
E = 2.00 Å
F = 2.04 Å

(a)   (b)   (d)

Figure 17.5. The crystal structure of the complex salt [Bu₄N][L···F]·MeCN·CH₂Cl₂ (L = **42**).[137] C–H hydrogen atoms, tetra-*n*-butylammonium countercation and solvate molecules have been omitted for clarity. (a) Red dashed lines indicate the six H-bonds donated by the six amide N–H fragments to the well-encapsulated fluoride ion; (b) view of the complex along the trigonal axis joining the two tertiary amine nitrogen atom of the amide cryptand (N_{tert}---N_{tert} distance: 7.39 Å); (c) triangles obtained by linking the amide hydrogen atoms belonging to the same tren subunit, top view: the torsional angle between the two triangles is ~36°, intermediate between that for a trigonal prism (0°) and that for an octahedron (60°); (d) lateral view.

Fluoride is well included into the amide cryptand cavity, positioned at an equal distance from the tertiary amine nitrogen atoms (3.68 and 3.71 Å). It receives six H-bonds from the six amide hydrogen atoms with an average N–H···F distance of 2.01 ± 0.03 Å. In Figure 17.5(c), the hydrogen atoms of the N–H fragments of each tren subunit have been linked to give a nearly equilateral triangle: the two triangles form a twist angle $\theta \sim 36°$, which indicates a coordination polyhedron of the H-bonds slightly closer to the octahedron ($\theta = 60°$) than to the trigonal prism ($\theta = 0°$).

Figure 17.6 shows the crystal structure of the complex salt [Bu₄N][L···F]·CHCl₃·CH₃COOC₂H₅ (L = **41**).[137] The structure is similar to that of the corresponding fluoride complex of the amide cryptand **42**. In particular, F⁻ receives six H-bonds from the six N–H fragments of the receptor. However, the average N–H···F distances (2.18 ± 0.06 Å) are distinctly higher than those observed in the complex of **42**. This is due to the fact that F⁻ receives three additional hydrogen bonds from the three facing

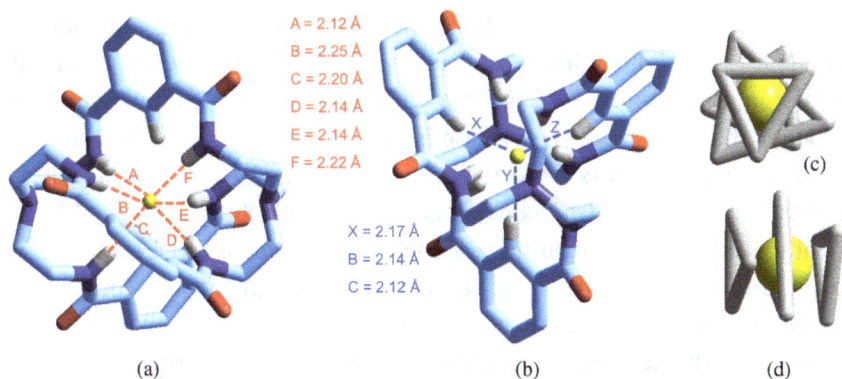

Figure 17.6. The crystal structure of the complex salt [Bu₄N][L···F]·CHCl₃· CH₃COOC₂H₅ (L = **41**).[138] C–H hydrogen atoms, tetra-*n*-butylammonium countercation and solvate molecules of chloroform and ethylacetate have been omitted for clarity. (a) Red dashed lines indicate the six H-bonds donated by the six amide N–H fragments to the well-encapsulated fluoride ion; (b) a different view of the complex which emphasises the three H-bonds donated to F⁻ by the three facing aromatic C–H fragments, blue dashed lines; on the whole, F⁻ receives nine H-bonds from the receptor; (c) the two triangles obtained by linking the amide hydrogen atoms belonging to the same tren subunit and the triangle in the middle, obtained by linking the three aromatic C–H hydrogens involved in H-bonds to F⁻, top view; (d) lateral view.

C–H fragments of the three 1,3-xylyl spacers (average C–H···F distance 2.14 ± 0.03 Å).

The stability in solution of these complexes has been evaluated by performing ¹H NMR titrations in MeCN. Studies were extended to chloride and bromide. Pertinent log *K* values are reported in Table 17.1.

Table 17.1. Log *K* values for the equilibrium: L + X⁻ ⇋ [L···X]+ in MeCN (L = **44**, **45**).[136]

Receptor L	F⁻	Cl⁻	Br⁻
41	4.5	2.9	<1
42	>5	3.5	1.6

For both amide cryptands, the stability of the halide complexes decreases along the series F⁻ > Cl⁻ > Br⁻, which reflects the decreasing basicity of the anion. Then, it may be surprising that the fluoride complex of **42** (six H-bonds) is more stable than the corresponding complex of **41** (nine H-bonds). In this regard, it must be noted that the six N–H⋯F bonds in the complex of **42** are distinctly shorter and presumably stronger than the six N–H⋯F bonds of complex **41**. It is true that the latter complex benefits from three additional C–H⋯F bonds, but it is also true that the C–H fragment is much less polarised than the N–H fragment and its H-bond donating tendencies are expected to be remarkably lower. In conclusion, the three aromatic C–H fragments of receptor **41** intend to cooperate to ensure the stability of the fluoride complex, but ultimately disturb the H-bond donating tendencies of the intrinsically more effective amide N–H fragments. It is possible that a similar mechanism determines the lower stability of the chloride and bromide complexes of **41** compared to **42**.

All the cryptands considered until now result from the proper linking of two tripodal scaffolds. Linking of two tren moieties with three spacers yielded bistren in the amine or amide version (Figure 17.7(a)). Bowman-James and coworkers decided to use another branched polyamine as a scaffold, the hexamine *penten* (*N,N,N′,N′*-tetrakis(2-aminoethyl) ethane-1,2-diamine).[139] Penten (the name was derived from pentaethylenehexamine) was synthesised in 1952 by Gerold Schwarzenbach, ETH, Zurich, as a full-amine analogue of EDTA.[140] However, penten did not enjoy the same success in coordination chemistry as a ligand of the other branched polyamine tren. Linking of two penten subunits gives rise to a macrotricyclic cryptand, as illustrated in Figure 17.7(b).

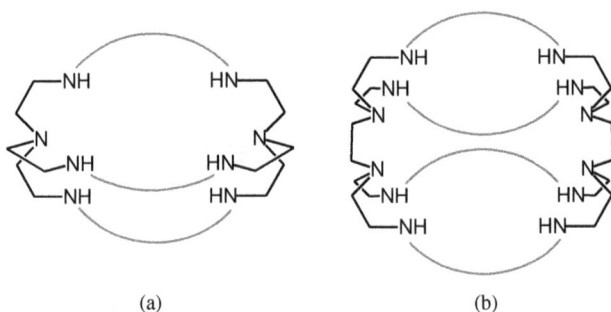

(a)   (b)

Figure 17.7. The generation of macrobicyclic (a) and macrotricyclic (b) cryptands. (a) originated by linking of two *tren* subunits by three spacers; (b) originated by linking of two *penten* subunits by four spacers.

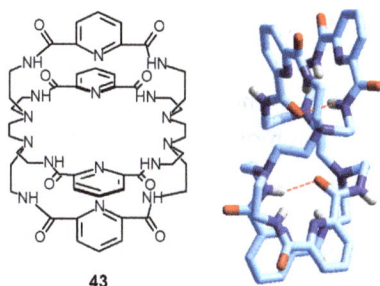

43

Figure 17.8. The crystal structure of the bispenten cryptand **43**.[139] C–H hydrogen atoms and the six solvating water molecules have been omitted for clarity. Red dashed lines indicate intramolecular hydrogen bonds: both N–H···O distances are 2.11 Å.

Through a multistep synthesis, Bowman-James obtained the bispenten amide cryptand **43** with an overall yield of 3%.[139] The crystal structure of **43** is shown in Figure 17.8. The receptor cavity is narrow and elongated due to two intramolecular hydrogen bonds and two π–π interactions involving the pyridine subunits of the spacers. Such an oblong space seems to be designed for linear polyatomic anions. Indeed, the bispenten amide receptor was found to form stable complexes with the linear triatomic anions HF_2^- and N_3^-. The structures of the corresponding salts $[Bu_4N][L \cdots HF_2] \cdot 3H_2O$[139] and $[Bu_4N][L \cdots N_3] \cdot 3H_2O$[141] are shown in Figure 17.9.

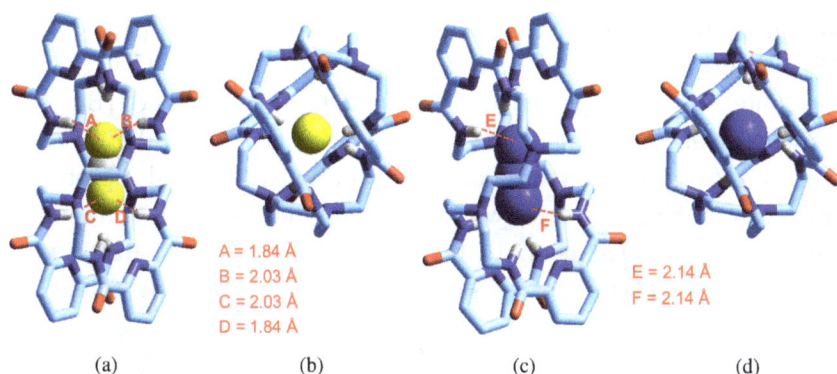

A = 1.84 Å
B = 2.03 Å
C = 2.03 Å
D = 1.84 Å

E = 2.14 Å
F = 2.14 Å

(a)   (b)   (c)   (d)

Figure 17.9. The crystal structures of the complex salts (a) $[Bu_4N][L \cdots HF_2] \cdot 3H_2O$, (b) top view, and (c) $[Bu_4N][L \cdots N_3] \cdot 3H_2O$, (d) top view, (L = **43**).[141] C–H hydrogen atoms, tetra-*n*-butylammonium counterions and solvating water molecules have been omitted for clarity. Red dashed lines indicate intramolecular hydrogen bonds involving N–H fragments and anion terminal atoms.

Figure 17.9(a) shows that HF_2^- is well included in the receptor's cavity. In particular, the structures of the cryptand framework in the absence (Figure 17.8) and in the presence of HF_2^- are very similar and superimposable. Thus, the inclusion of anion does not induce any serious conformational modification, apart from the ~180° rotation of two amide groups, which involves the breaking of two intramolecular N–H···O hydrogen bonds. However, the π–π interactions are maintained.

In the complex, for each terminal fluorine atom, the anion receives two H-bonds from two facing N–H fragments, for a total of four. These bonds bring the F---F distance to 2.48 Å, i.e. ~10% larger than that of the isolated, unperturbed HF_2^-,[142] a behaviour reflecting the 'elastic' nature of the two-electron three-centre F–H–F bond.

Azide forms an inclusion complex with the amide cryptand **43**, which is almost isostructural with that of hydrogen difluoride, whose structure is shown in Figure 17.9(b).[141] Each terminal nitrogen atom of N_3^- receives one H-bond from a facing N–H fragment for a total of two, which may lead to a lower stability compared to HF_2^-. Indeed, on ^{1}H NMR titration of a solution of **43** in DMSO-d^6 with $[Bu_4N]HF_2$ and $[Bu_4N]N_3$, the inclusion constants were determined: HF_2^-, log K = 3.74; N_3^-, log K = 2.53.[137] The stability sequence reflects the number of hydrogen bonds present in each complex.

Finally, it has to be mentioned that the synthesis of the hydrogen difluoride complex was serendipitous. In fact, crystals of $[Bu_4N][L···HF_2] \cdot 3H_2O$ salt were obtained on evaporation of an MeCN/CHCl$_3$ solution containing the bispenten amide cryptand and an excess of $[Bu_4N]F$, a process aimed to obtain the fluoride complex.[139] The HF_2^- anion, hydrogen difluoride (often named, less correctly, bifluoride) probably formed from the reaction of two F$^-$ ions with one amide N–H fragment acting as a Brønsted acid. In fact, fluoride typically interacts with amides, ureas and thioureas (LH)[143–145] according to the two following stepwise equilibria:

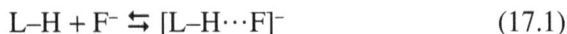

$$L–H + F^- \leftrightarrows [L–H···F]^- \qquad (17.1)$$

$$[L–H···F]^- + F^- \leftrightarrows L^- + HF_2^- \qquad (17.2)$$

In the first step (17.1), a classic hydrogen bond complex is formed: [L–H···F]⁻. Then, in the second step (17.2), deprotonation of the amide N–H fragment takes place with formation of the HF_2^- ion. The stepwise process referring to a fragment of the amide cryptand **43** is pictorially illustrated in Figure 17.10.

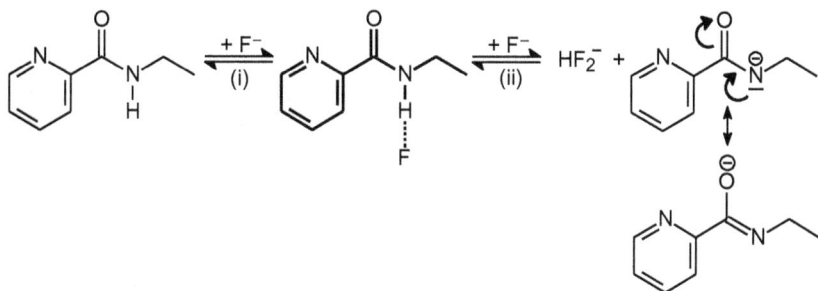

Figure 17.10. The interaction of fluoride with an amide group: (i) formation of the H-bond complex; (ii) deprotonation of the N–H fragment and formation of hydrogen difluoride. The deprotonated amide subunit is stabilised by π delocalisation.

Step (ii) is driven by the extraordinary stability of HF_2^-, the most stable of hydrogen bonding complexes.[146] Thus, *one* F⁻ is a base of moderate strength, but *two* F⁻ are a very strong base.[147]

18

The Taming of Peroxide (By a Bistren Amide Cryptand)

Peroxide, O_2^{2-}, is an intrinsically unstable anion. Its poor stability is because of the repulsion between the 3 + 3 lone pairs present in the two oxygen atoms. Na_2O_2 is stable as a solid, but reacts violently with water to give hydrogen peroxide and sodium hydroxide:

$$Na_2O_2(s) + 2H_2O(aq) \rightarrow H_2O_2(aq) + 2Na^+(aq) + 2OH^-(aq)$$

Ionic peroxides are insoluble in organic media and, in a suspension, tend to oxidise the solvent. Organometallic chemistry has demonstrated that molecular oxygen can be safely and reversibly reduced to peroxide by benefiting from coordination to a given metal centre. The first example was given by the Vaska's complex, $[Ir^ICl(CO)(PPh_3)_2]$ (**44**).[148] $[Ir^ICl(CO)(PPh_3)_2]$ is a versatile square planar diamagnetic complex of iridium(I), which, in benzene solution, is able to uptake a variety of substrates, including dioxygen. The geometrical details of the reaction of Vaska's complex with dioxygen are illustrated in Figure 18.1. Ir^I releases two electrons to O_2 to give O_2^{2-}, which binds Ir^{III} in a chelating mode. A diamagnetic 18-electrons complex forms, with a distorted octahedral

183

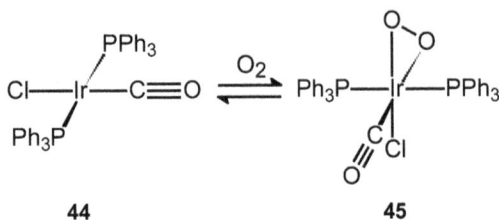

Figure 18.1. Vaska's complex (**44**) [Ir^{I}Cl(CO)(PPh$_3$)$_2$], square geometry, uptakes a molecule of O$_2$, which is reduced to O$_2^{2-}$ by Ir^I ($\rightarrow$ Ir^{III}). O$_2^{2-}$ binds Ir^{III} in a chelating mode. A complex of distorted octahedral geometry forms (**45**).

geometry. The process is reversible. In particular, the Ir^{III} complex loses dioxygen and reverts to the parent Ir^I four-coordinate complex on heating or purging the solution with an inert gas. The process can be visually monitored by a neat colour change: from orange (Ir^{III}) back to yellow (Ir^I).

[Ir^{I}Cl(CO)(PPh$_3$)$_2$] can therefore be considered an artificial oxygen carrier. Peroxide is bound to Ir^{III} by both the oxygen atoms, in the aptly named *side-on* bonding mode. In the natural carriers myoglobin and haemoglobin, by contrast, O$_2$ binds *end-on*, attaching to the metal through only one oxygen atom. Vaska was the first to demonstrate that the otherwise reactive and intractable peroxide ion can be made dormant and compliant by metal coordination.

Daniel Nocera, MIT, Cambridge, Massachusetts, considered the opportunity to stabilise O$_2^{2-}$ in a solution by the inclusion of a bistren amide cryptand, thus profiting from hydrogen bonding interactions. He did not start from peroxide (O$_2^{2-}$), but from superoxide (O$_2^-$). In particular, on treating a slurry of the *tert*-butyl-substituted hexacarboxamide cryptand **46** with 2.2 equivalent of KO$_2$ in DMF, dioxygen bubbled out from the solution and an inclusion complex of peroxide was formed with a yield of 74%.[149] The crystal structure of the complex is shown in Figure 18.2.[150] The peroxide ion is well inserted into the receptor's cavity. Each peroxide oxygen atom receives three H-bonds from the amide fragments of the facing tren subunit (N–H···O distances ranging from 1.88 to 1.93 Å). The amide hydrogen atoms are positioned at the corners of a polyhedron with a twist angle $\theta = 38 \pm 1°$,

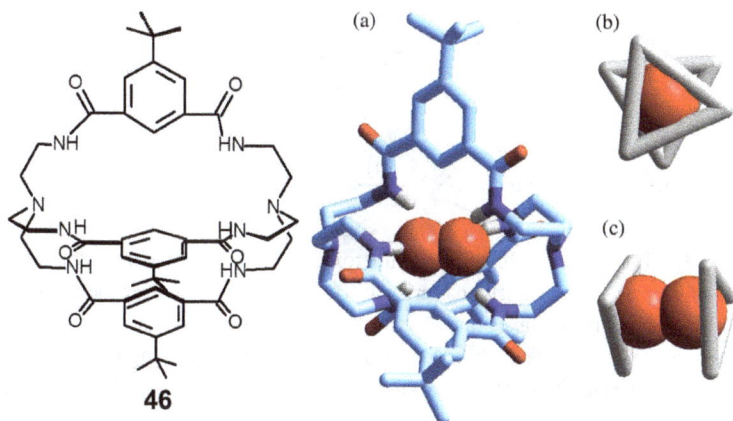

Figure 18.2. (a) The crystal structure of $K_2[L\cdots O_2]\cdot 5DMF$ (L = **46**)[150]; C–H hydrogen atoms, potassium ions and solvating DMF molecules have been omitted for clarity; each oxygen atom of the included peroxide receives three H-bonds from the three N–H fragments of the facing tren subunit; the N_{tert}---N_{tert} distance is 8.18 Å; (b) the two triangles obtained by linking the amide hydrogen atoms belonging to the same tren subunit, which form an average torsional angle $\theta = 38 \pm 1°$; (c) lateral view.

intermediate between that of a trigonal prism ($\theta = 0°$) and that of an octahedron ($\theta = 60°$). The presence of six H-bonds imparts particular stability to the peroxide cryptate complex, which accounts for the occurrence of redox process (18.1), in which the superoxide, in the presence of the amide cryptand **46** (L), disproportionates to give dioxygen and the peroxide complex:

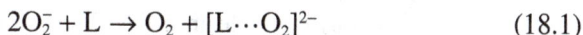

$$2O_2^- + L \rightarrow O_2 + [L\cdots O_2]^{2-} \qquad (18.1)$$

On the basis of cyclic voltammetry experiments, Nocera and coworkers demonstrated that process (18.1) occurs through the steps illustrated in Figure 18.3. In step (i), O_2^- is sequestered by the bistren amide cryptand **46** (L) to give the cryptate complex $[L\cdots O_2]^-$; then, in step (ii), the complex diffuses in solution and collides with an uncomplexed superoxide ion. This collision is much more probable than collision with another $[L\cdots O_2]^-$, because the voluminous cryptate complex is less mobile than the 'naked' anion O_2^-. At this point, the

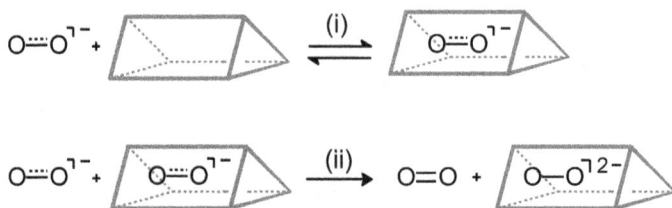

Figure 18.3. The disproportionation of superoxide in the presence of the bistren amide cryptand **46** (represented as a trigonal prism): (i) O_2^- is encapsulated by the cryptand; (ii) O_2^- in the cryptate complex takes one electron from a colliding uncomplexed O_2^- to give a more stable peroxide complex, while dioxygen develops.[149] The driving force of the entire process is the high stability of the peroxide cryptate complex (O_2^{2-} fixed to the cryptate by six H-bonds).

complexed O_2^- finds it convenient to take one electron from the uncomplexed O_2^-, transforming itself into the peroxide complex. In fact, O_2^{2-} forms a cryptate complex much more stable than O_2^- because, in view of its double negative charge, it is a more avid receiver of H-bonds.

Disproportionation of the toxic O_2^- radical in aerobic organisms is a task accomplished by superoxide dismutases (SOD). In particular, (Cu, Zn)-SOD is an enzyme containing copper(II) and zinc(II), which operates through the Cu^{II}/Cu^I redox couple: Cu^{II}-SOD oxidises O_2^- to O_2 and Cu^I-SOD which forms reduces another O_2^- to H_2O_2.[151] Cryptand **46** deactivates the harmful O_2^- ion through a more simple mechanism based on hydrogen bonding interactions. However, it operates on a stoichiometric basis, not catalytically, as SOD does.

$[L \cdots O_2]^{2-}$ (L = **46**) is stable in a DMF solution at room temperature for at least one month and remains intact even after heating for 100 minutes at 50°C, a feature which provides the opportunity to investigate the reactivity of the peroxide ion in solution, for instance towards carbon monoxide. The elimination of the toxic CO gas originating from industrial waste streams and vehicle exhaust is an important issue in environmental chemistry. A convenient process involves the reaction of CO with O_2^{2-} to give CO_3^{2-}. Such a process is typically mediated by a metal centre, as illustrated in Figure 18.4(a)[152]

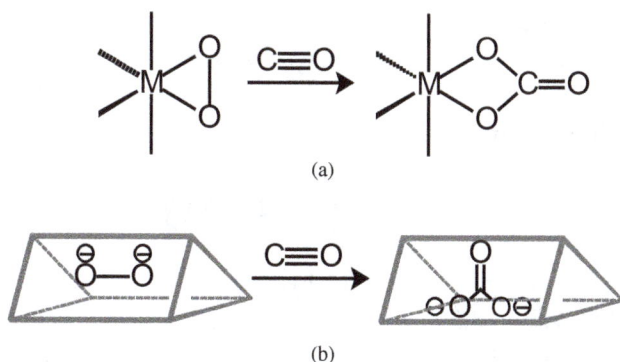

(a)

(b)

Figure 18.4. The oxidation in solution of CO to CO_3^{2-} by O_2^{2-}: (a) metal-driven reaction of metal-bound peroxide with carbon monoxide to give metal-bound carbonate; (b) the reaction of peroxide, encapsulated by cryptand **46**, with carbon monoxide to give carbonate.

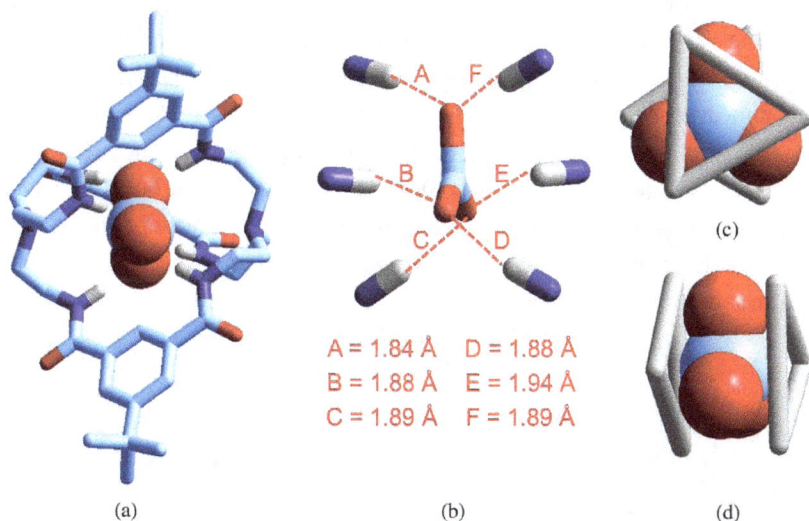

A = 1.84 Å   D = 1.88 Å
B = 1.88 Å   E = 1.94 Å
C = 1.89 Å   F = 1.89 Å

(a)   (b)   (c)

(d)

Figure 18.5. (a) The crystal structure of the complex salt $[Bu_4N]_2[L\cdots CO_3]$ (L = **46**);[153] C–H hydrogens and tetra-*n*-butylammonium ions have been omitted for clarity; the $N_{tert}\cdots N_{tert}$ distance is 7.69 Å; (b) network of hydrogen bonds involving amide N–H fragments and carbonate oxygen atoms, with N–H$\cdots$O distances typical of moderate H-bonds; (c) triangles obtained by linking the amide hydrogen atoms belonging to the same tren subunit, forming an average torsional angle $\theta = 52 \pm 3°$ and sandwiching a carbonate ion; (d) lateral view.

Nocera and coworkers observed that peroxide encapsulated in the hexacarboxamide cryptand **46** reacts with carbon monoxide in organic solvents at 40°C to form quantitatively the corresponding carbonate complex.[153] Such a complex has been isolated as a tetra-*n*-butylammonium salt. The crystal structure of the complex salt $[Bu_4N]_2[L\cdots CO_3]$ (L = **46**) is shown in Figure 18.5(a).[153] The triangular CO_3^{2-} is well included in the cavity of the bistren amide cryptand and is parallel to the triangles obtained by linking the amide hydrogens of each tren subunit (Figures 18.5(c) and 18.5(d)). Each oxygen of the carbonate receives two H-bonds from two N–H fragments belonging to different tren subunits (N–H$\cdots$O distances in Figure 18.5(b). The network of six N–H$\cdots$O hydrogen bonds makes the carbonate complex extremely stable, which accounts for the spontaneous occurrence of the CO oxidation. Labelling studies and ^{17}O solid-state NMR data confirm that two-thirds of the oxygen atoms in the encapsulated carbonate derive from peroxide dianion, while the carbon is derived from CO. These evidences indicate that CO enters the cryptand's cavity and goes on to react with the encapsulated O_2^{2-}, and excludes the reaction of CO in solution with an uncomplexed peroxide and subsequent replacement by carbonate of the included O_2^{2-}. This study has demonstrated that O_2^{2-} can oxidise CO without the assistance of a transition metal, if hosted and probably activated by a bistren hexamide cryptand. *Peroxidum agit quia fixatum*, one could say, by paraphrasing the well-known statement by Paul Ehrlich.

19

Coordination Chemistry of the Smallest Bistren Cryptand

Bistren amine cryptands have been considered in the previous Chapters as dinucleating ligands: bicyclic polyamines capable of hosting two metal centres; the length of the spacers determined the intermetallic distance and the space for a bridging anion, either monoatomic or polyatomic. We have not considered yet the 'smallest' bistren cryptand, **47** (bistren-C$_2$), in which the two tren subunit are linked by –CH$_2$CH$_2$– spacers. Bistren-C$_2$

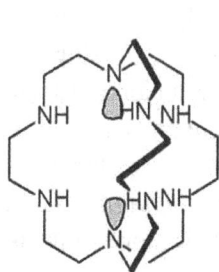
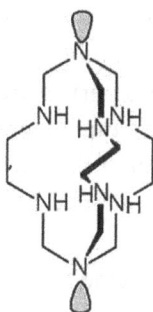
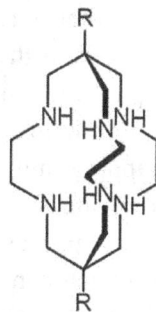

47	48	49
bistren-C$_2$	sepulchrand	sarcophagine

189

can be obtained through the 2 + 3 Schiff base condensation, from tren and glyoxal. The essential element for a good yield (>50%) appears to be the slow addition of glyoxal to tren at low temperatures (0°C).[154]

Bistren-C_2 is an octadentate ligand, but it possesses six secondary amine groups positioned at the corners of a trigonal prism/octahedron suitable for the coordination of a metal ion, transition and post-transition. In this sense, bistren-C_2 is reminiscent of the six-coordinating bicyclic octamine **48**, sepulchrand.[155] However, there exist substantial differences between the two ligands: (i) bistren-C_2 has two $N(CH_2CH_2-)_3$ caps and is expected to provide a larger cavity than sepulchrand, which possesses two $N(CH_2-)_3$ caps; (ii) the lone pair of the tertiary nitrogen atom of bistren-C_2 points inside the cavity and can offer alternative/additional bonding to the included metal, whereas the lone pair of sepulchrand points outside; (iii) synthesis of bistren-C_2 complexes takes place classically, by mixing solutions of the ligand and of the metal salt, synthesis of sepulchrate complexes strictly requires a *template* effect by a d^6 low-spin metal ion (e.g. Co^{III}), which remains permanently imprisoned inside the ligating framework. Moreover, demetallation of the sepulchrate complex causes decomposition of the ligand, due to the intrinsic instability of the aminal groups of the caps. Alan Sargeson, University of Canberra, brought to coordination chemistry the fascinating ligand **48**, which he christened *sepulchrand* (and sepulchrate its metal complexes),[155] inspired by the solemn and slightly funereal fashion of naming ligands introduced by Lehn with cryptands and cryptates. There is an obvious lexical difference between crypt and sepulchre, which is maintained in their chemical counterparts: holy relics can be periodically removed from the crypt and exposed to the worship of believers (labile complex); remains of a corpse are never taken out from the sepulchre (inert complex). Later on, Sargeson, using CH_3NO_2 instead of NH_3 as a capping unit, obtained the Co^{III} complex of the hexamine **49**, named, along the same lugubrious line, *sarcophagine* (a polyamine exerting the properties of a sarcophagus).[156] Sarcophagine is a stable polyamine, which can be obtained as a free ligand through the rather elaborate demetallation of its Co^{III} complex.[157] Sarcophagine (R = NO_2, NH_2 or H) has been made to react with a variety of metals to give complexes with interesting structural properties.

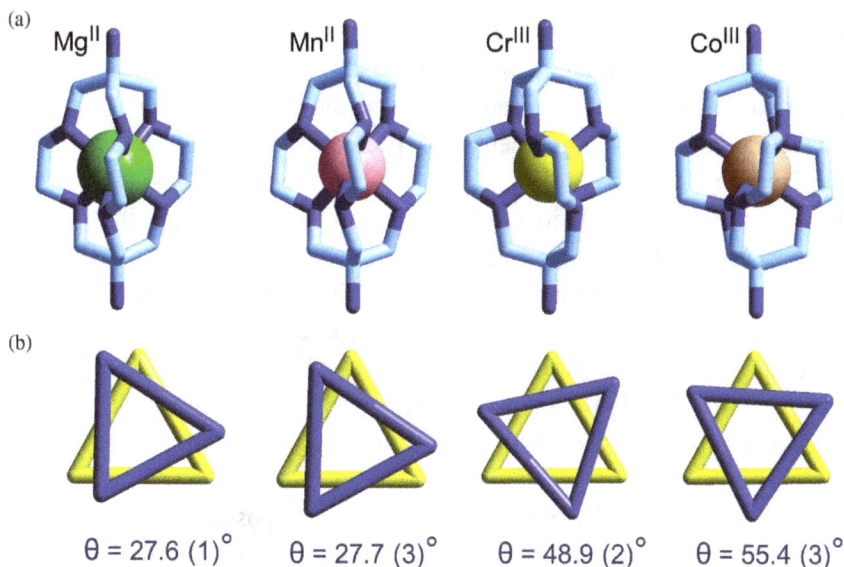

Figure 19.1. (a) Structures of metal complexes of sarcophagines (sar), as obtained from X-ray diffraction studies; Cr^{III} is complexed by sar with R = NH_2; all other metals are complexed by sar with R = NH_3^+; (b) triangles obtained by joining the nitrogen atoms of the secondary amine of each C(CH₃NH-)₃ cap; θ = twist angle for each pair of triangles. Complex salts are: [Mg^{II}(sar-NH₂-NH₃)](NO₃)₃ · H₂O,[157] [Mn^{II}(sar-2NH₃)](NO₃)₄ · H₂O,[158] [Ni^{II}(sar-2NH₃)]Cl₄ · H₂O,[159] [Co^{III}(sar-2NH₃)]Cl₅ · 2H₂O.[160] Hydrogen atoms, counteranions and solvating water molecules have been omitted for clarity.

Some examples are shown in Figure 19.1.(a). Sarcophagine complexes show coordination geometries intermediate between the trigonal prism and the octahedron, with the degree of distortion changing with the electronic configuration of the metal ion. In the Mg^{II} complex, the twist angle θ (= $27.6 \pm 0.1°$) is midway between that of the trigonal prism ($\theta = 0°$) and that of the octahedron ($\theta = 60°$). Mg^{II} is a spherical ion that does not present any geometrical preference. This suggests that the coordinated sarcophagine framework, in the Mg^{II} complex, is in its most stable conformation, from a purely sterical point of view. Crystal field theory states that for a transition metal ion (electronic configuration d^n, n = 1–9) octahedral coordination is energetically more favourable than trigonal prismatic. For instance, a d^3/d^8 metal experiences a crystal field stabilisation energy CFSE = −12 Dq in an octahedral geometry and CFSE = −10.72 Dq in a trigonal prismatic arrangement. This advantage of having -1.28 Dq explains why in the Cr^{III} (d^3) complex the twist angle ($\theta = 48.9 \pm 0.2°$) has moved to a value much closer to that of the octahedron. However, the greatest advantage of the octahedral coordination geometry is expected for a low-spin d^6 electronic configuration, that of Co^{III} (CFSE = -24 Dq, versus -21.44 Dq for the trigonal prism). Indeed, in the Co^{III} sarcophagine complex, the closest value to $60°$ is observed ($\theta = 55.4 \pm 0.3°$), which makes the coordination geometry nearly octahedral. On the other hand, the Mn^{II} metal centre, high-spin d^5, has a CFSE = 0 Dq in any coordination geometry and its sarcophagine complex shows a twist value $\theta = 27.7 \pm 0.3°$, the same value observed for the Mg^{II} complex, which corresponds to the less strained conformational arrangement of the ligating framework, in the absence of any crystal field effect.

Bistren-C_2 is similar to sarcophagine in that it possesses a set of three ethylenediamine fragments, suitable for metal coordination. However, similarity is only formal, and metal complexes of **47** display rather different properties and reactivity. The main differences are: (i) the two tertiary nitrogen atoms of tren subunits (exhibiting an *in,in* configuration) play an active role in the coordination of the included metal; (ii) the two tren subunits impart to bistren-C_2 a flexibility absent in sarcophagine derivatives.

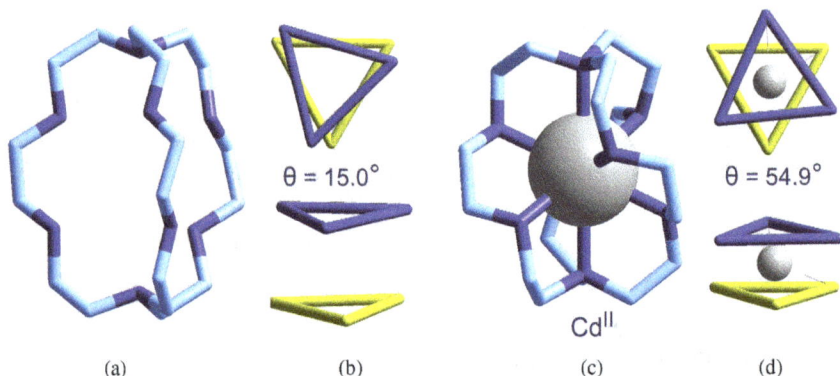

$\theta = 15.0°$ Cd^{II} $\theta = 54.9°$

(a)   (b)   (c)   (d)

Figure 19.2. Crystal structure of bistren-C_2, L·18H$_2$O (L = **47**)[154] (a) and of its cadmium complex salt [Cd^{II}(L)](BF$_4$)$_2$;[161] (b); triangles obtained by linking the secondary amine nitrogen atoms of each tren subunit of the uncomplexed ligand (c) and of its Cd^{II} complex (d), with average values of torsion angles θ. Hydrogen atoms, solvating water molecules and tetrafluoroborate counteranions have been omitted for clarity.

Figure 19.2 compares the crystal structures of the uncomplexed bistren-C_2 molecule (a)[154] and its complex with Cd^{II} (c).[161] The empty cryptand presents an ellipsoidal cavity, with a N_{tert}---N_{tert} distance of 6.37 Å. The spherical Cd^{II} ion ($4d^{10}$ electronic configuration) is bound to all the eight amine groups (distance Cd^{II}–NH(sec): 2.52 Å; Cd^{II}–N(tert): 2.72 Å). In order to grant full amine coordination, in particular bonding of the two tertiary amine nitrogen atoms, the ligand contracts its cavity, by reducing the N_{tert}---N_{tert} distance to 5.55 Å. To do that, it makes use of a well-established spring and screw mechanism, which involves the rotation by 40° of the triangle defined by the secondary amine nitrogen atoms of a tren subunit, with respect to triangle of the other tren subunit. Thus, the increase of the twist angle from 15° to 55° does not result from a search of any CFSE advantage (to which the post-transition metal ion Cd^{II} is insensitive), but it is only a way to favour the metal coordination of the two tertiary nitrogen atoms. In the corresponding cadmium(II) sarcophagine complex,[160] the metal is coordinated by the six secondary amine nitrogen atoms (average Cd^{II}–N distance 2.30 ± 0.03 Å) with a twist angle $\theta = 27.3 \pm 0.3°$, the value expected for d^{10} metal complexes which do not profit from a CFSE contribution. Thus, in both $[Cd^{II}(sar\text{-}2NH_3)]^{4+}$ and $[Cd^{II}(bistren\text{-}C_2]^{2+}$, the metal fully profits from the binding potential of the ligand, establishing coordinative interactions with all the available donor atoms, six and eight, respectively.

Things go differently with transition metals. Titration experiments in aqueous solution have been carried out in order to investigate the interaction of Cu^{II} with bistren-C_2.[162] Figure 19.3(a) shows the concentration profiles of the species present at the equilibrium moving from pH 2 to pH 12 (Cu^{II}/bistrenC$_2$ ratio 1:1). In the still more acidic region, at the beginning of the titration, a stable complex species forms, $[Cu^{II}(LH)]^{3+}$, which predominates over the pH interval 3–9. Then, the $[Cu^{II}(L)]^{2+}$ complex prevails, which reaches 100% at a definitely alkaline pH: 12.

Figure 19.3(b) shows the crystal structure of the $[Cu^{II}(L)]^{2+}$ complex.[163] In contrast to what is observed for the corresponding Cd^{II} complex, the metal neither occupies the central position of the cavity, nor is coordinated to all available nitrogen atoms. It is bound to four nitrogen atoms of one tren subunit and to a secondary amine nitrogen atom of the

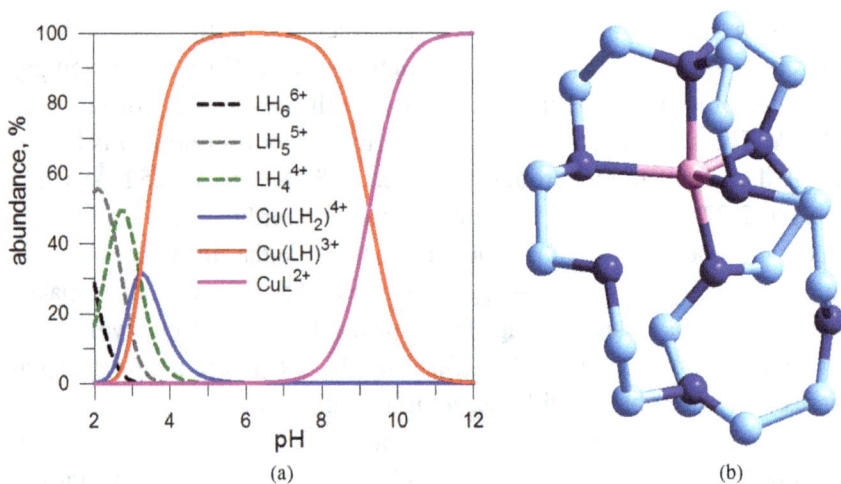

(a)   (b)

Figure 19.3. (a) Percent concentration profiles of the species present at the equilibrium over the course of the titration with standard NaOH of a solution 10^{-3} M of bistren cryptand **47** (L) and 10^{-3} M in $Cu^{II}(CF_3SO_3)_2$, containing excess acid[162]; (b) crystal structure of the $[Cu^{II}(bistren-C_2)](BPh_4)_2$ complex salt[163]; hydrogen atoms and tetraphenylborate counteranions have been omitted for clarity.

other tren subunit, for a total of five, according to a distorted trigonal bipyramidal geometry. Copper(II) exhibits a pronounced preference for five-coordination. It is possible that coordination to the six secondary amine nitrogen atoms to give a tris-ethylenediamine-type complex, observed with cadmium, would require a very endergonic conformational rearrangement of the cryptand framework.

A crystal structure of the $[Cu^{II}(LH)]^{3+}$ (L = **47**) does not exist. However there exists a structure of the corresponding complex of the *N*-tetramethyl derivative **50**, which is shown in Figure 19.4.[164] The copper(II) ion is coordinated by the four nitrogen atoms of one tren subunit, while an oxygen atom occupies the other axial position of a trigonal bipyramid. The nature of the metal-bound oxygen atom deserves detailed consideration. Three hypothetical arrangements are illustrated in Figure 19.5. The most obvious hypothesis would be to consider that a water molecule is present in the cavity of the monoprotonated cryptand LH^+. This water molecule on one side coordinates Cu^{II}, as it may happen

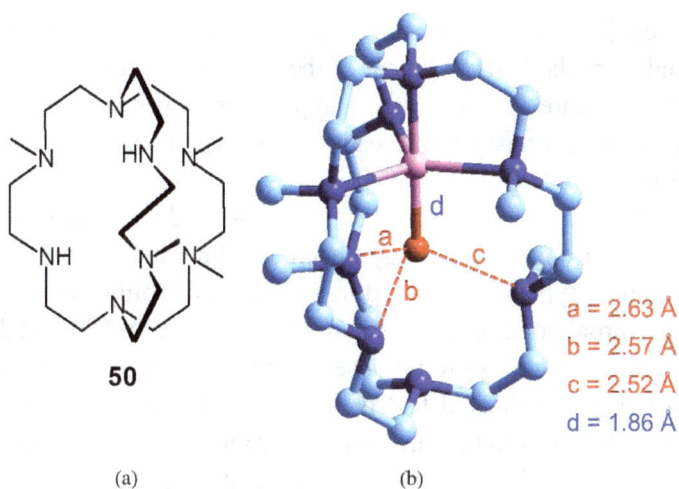

(a)   (b)

Figure 19.4. The crystal structure of the complex salt $[Cu^{II}(L\cdots H_2O](NO_3)_3 \cdot MeCN$ (L = **50**).[164] Hydrogen atoms, nitrate counterions and solvate MeCN molecule have been omitted for clarity. The unusually short Cu^{II}–O and N---O distances suggest that the complex should be described better by the formula $[Cu^{II}(LH_3)(O)]^{3+}$: an O^{2-} ion is coordinated to Cu^{II} and receives three H-bonds from three secondary ammonium groups of the lower tri-protonated tren subunit.

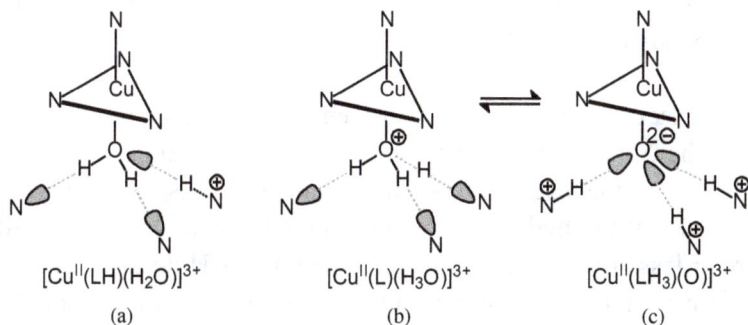

$[Cu^{II}(LH)(H_2O)]^{3+}$ $[Cu^{II}(L)(H_3O)]^{3+}$ $[Cu^{II}(LH_3)(O)]^{3+}$

(a)   (b)   (c)

Figure 19.5. Possible coordination modes of the oxygen atom in the complex of Cu^{II} with LH^+ (L = bistren-C_2, **47** or **50**): (a) an included water molecule is bound to Cu^{II} and one of the oxygen lone pairs receives an H-bond from a secondary ammonium groups of the lower tren subunit; (b) an oxonium ion is included: its oxygen atom coordinates Cu^{II}, while its three polarised hydrogen atoms donate three H-bonds to the three secondary nitrogen atoms of the lower tren subunit; (c) O^{2-} is bound to the metal and receives three H-bonds from three N–H fragments of the secondary ammonium groups of the lower tren subunit.

in a classical $[Cu^{II}(tren)H_2O]^{2+}$ complex, while, on the other side, receives an H-bond from the N–H fragment of the ammonium group of the lower tren subunit (Figure 19.5(a)). Such a hypothesis, however, is not consistent with some peculiar structural features of the complex. In particular, the O---N distances involving the three secondary nitrogen atoms of the lower tren subunit are unusually short (2.57 ± 0.05 Å), indicating the formation of especially strong O–H$\cdots$N hydrogen bonding interactions: in fact, O---N distances typically observed in the inclusion complexes of oxoanions in hexaprotonated bistren receptors range between 2.80 and 3.00 Å. This suggested the authors to propose as a preliminary hypothesis that an oxonium ion is coordinated to Cu^{II} and its highly polarised hydrogens donate three H-bonds to the three secondary amine nitrogen atoms of the lower tren subunit (Figure 19.5(b)). However, it is questionable that H_3O^+ can behave as a ligand, and, in any case, it should establish a rather weak coordinative interaction with a metal. On the contrary, the Cu^{II}–O distance is unusually short: 1.86 Å. Notice that in the $[Cu^{II}(tren)H_2O]^{2+}$ complex, such a distance is 2.01 Å,[165] in the $[Cu^{II}(trenBn_3)H_2O]^{2+}$ (Bn = benzyl) complex it is 1.96,[166] and in the dicopper(II) complex of cryptand **37** it is 1.98 Å (see Figure 17.5).[128] This suggests that the bonding arrangement could be better described in Figure 19.5(c), in which an oxide ion is strongly coordinated to Cu^{II}. The metal-bound O^{2-} ion receives in addition three H-bonds from the three secondary ammonium groups of the lower tren subunit.

The exact position of the hydrogen atoms cannot be determined through the X-ray diffraction technique, which prevents from a sound assessment of the real bonding arrangement. It has been previously mentioned that the hydrogen bonding interaction can be considered as a more or less advanced proton transfer from a D–H donor to an acceptor A.[134] We can therefore suggest that the tautomeric equilibrium (b) $\leftrightarrows$ (c) in Figure 19.5 is highly, if not completely, displaced to the right hand. A synergistic effect seems to operate, by which the more advanced is the O-to-N proton transfer, the stronger is the Cu^{II}–O bond. In any case, the combination of the especially intense metal–ligand and hydrogen bonding interactions accounts for the high stability of the tripositive complex and for its dominating presence over an unusually broad pH interval (see Figure 19.3).

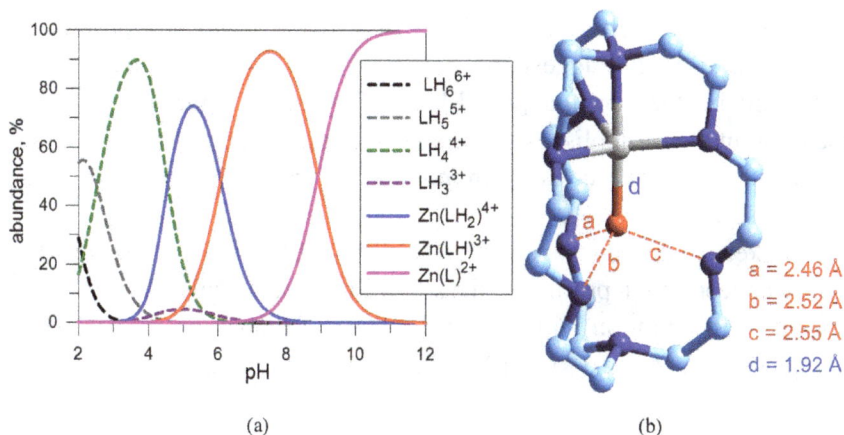

Figure 19.6. (a) Percent concentration profiles of the species present at the equilibrium over the course of the titration with standard NaOH of a solution 10^{-3} M of cryptand **47** (L) and 10^{-3} M in $Zn^{II}(CF_3SO_3)_2$, containing excess acid[162]; (b) crystal structure of $[Zn^{II}(L)(H_3O)](ClO_4)_2(BPh_4)$[164]; hydrogen atoms, perchlorate and tetraphenylborate counteranions have been omitted for clarity. The real structure should be better described by the formula $[Zn^{II}(LH_3)(O)](ClO_4)_2(BPh_4)$, in analogy with the Cu^{II} complex illustrated in Figure 19.4.

Zinc(II) displays quite a similar behaviour. Figure 19.6(a) shows the concentration profiles of the species at equilibrium over the course of a titration with standard base of an aqueous solution of 10^{-3} M both in bistren-C_2 and in $Zn(CF_3SO_3)_2$.[162] Zinc(II), $3d^{10}$, does not benefit from ligand field effects and forms less stable complexes than copper(II). Thus, complex species $[Zn^{II}(LH_2)]^{4+}$, $Zn^{II}(LH)]^{3+}$ and $Zn^{II}(L)]^{2+}$ form at higher pH values than observed for Cu^{II}. The tripositive complex, formally described by the formula $Zn^{II}(LH)]^{3+}$, is the dominating species over the 6–9 pH interval. The crystal structure of the $[Zn^{II}(L)(H_3O)](ClO_4)_2(BPh_4)$ complex salt is shown in Figure 19.6(b)[164] and presents a surprising analogy with the corresponding complex of Cu^{II}, shown in Figure 19.4. In fact, Zn^{II} is coordinated by the four amine nitrogen atoms of one tren subunit of bistren-C_2 and by one oxygen atom, according to a trigonal bipyramidal geometry. The Zn^{II}–O bond length (1.92 Å), distinctly lower than observed in the $[Zn^{II}(tren)(H_2O)]^{2+}$ complex (2.01 Å),[167] and the O---N distances (2.51 ± 0.05 Å), shorter than those typically observed in the inclusion complexes of oxoanios with hexaprotonated cryptands, strongly suggest the presence of the bonding arrangement described by the formula $[Zn^{II}(LH_3)(O)]^{3+}$ and in Figure 19.5(c): the metal-bound O^{2-} establishes especially strong hydrogen bonding interactions with the three secondary ammonium groups of the lower tren subunit of bistren-C_2.

Nickel(II) displays a completely different behaviour. Figure 19.7 shows concentration profiles of the species present at equilibrium over the course of the titration with standard NaOH of a solution of 10^{-3} M of cryptand **47** (L) and 10^{-3} M in $Ni^{II}(CF_3SO_3)_2$, containing excess acid.[162]

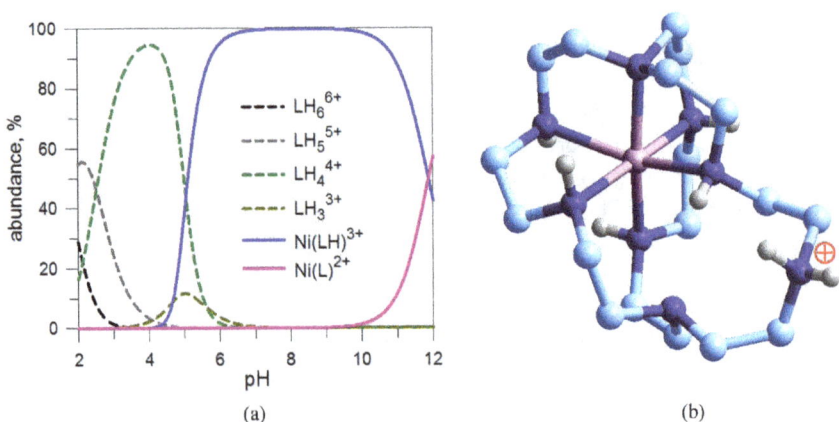

Figure 19.7. (a) Percent concentration profiles of the species present at the equilibrium over the course of the titration with standard NaOH of a solution 10^{-3} M of cryptand **47** (L) and 10^{-3} M in $Ni^{II}(CF_3SO_3)_2$, containing excess of triflic acid[162]; (b) crystal structure of $[Ni^{II}(LH)](ClO_4)_3 \cdot H_2O$;[162] C–H hydrogen atoms, perchlorate counteranions and solvating H_2O molecule have been omitted for clarity.

Also in this case, as observed for copper(II), the tripositive complex, $[Ni^{II}(LH)]^{3+}$, dominates, being present at 100% over the pH interval 6–10. Then, the $[Ni^{II}(L)]^{2+}$ complex forms. However, the crystal structure shown in Figure 19.7(b) shows that the relatively high stability of the complex does not result from the active presence of an oxide ion in the cavity, as observed for corresponding Cu^{II} and Zn^{II} complexes.[162] Instead, Ni^{II} is coordinated by six amine nitrogen atoms (four by one tren subunit, two from the other), according to a nearly regular octahedral geometry. The remaining secondary amine group is protonated. Ni^{II}, $3d^8$ high-spin, has a defined preference for octahedral coordination and, in the $[Ni^{II}(LH)]^{3+}$ complex, it seems to reach its beloved geometry without inducing serious constraints in the cryptand framework. On the other hand, the ammonium group is quite distant from Ni^{II} (N---Ni^{II} distance 5.07 Å), which should minimise electrostatic repulsion. This may explain the resistance of $[Ni^{II}(LH)]^{3+}$ to deprotonation and the delayed appearance of the $[Ni^{II}(L)]^{2+}$ complex. Proton release, in fact, occurs at the pH typically observed for a secondary amine ($pK_a = 10$–11), while Ni^{II} maintains its coordinative arrangement.

The different behaviour of the triposive complexes of bistren-C_2, Cu^{II} and Zn^{II} on one hand, Ni^{II} on the other, is a further demonstration of how electronic configuration determines the geometrical preferences of the envisaged metal ions. Cu^{II}, d^9, subject to Jahn–Teller distortions, is a coordinatively versatile cation ('plastic') and entirely profits from the unique opportunity offered by LH^+: full binding to one tren subunit, completion of five-coordination with a water molecule, rearrangement of the hydrogen bonding network, eventually leading to the O^{2-}/LH_3^{3+} system. Zn^{II}, d^{10}, as a 'spherical' cation, does not exhibit definite geometrical preferences: thus, it does not have any difficulty to take the path chosen by its close fellow, Cu^{II}. On the other hand, Ni^{II}, in the high-spin state, does not conceive any other coordination geometry than the octahedral one, which it wants to achieve at all costs, whether the cryptand is protonated or not.

Now, one could wonder whether bistren-C_2 could accommodate *two* metal ions in its cavity. On reaction of bistren-C_2 with excess of $Cu^{II}(NO_3)_2 \cdot 3H_2O$ in MeOH, an intense blue colour developed. On evaporation, a salt with the formula $[Cu_2(L)](NO_3)_3 \cdot 2H_2O$ precipitated in form of blue crystals. Such a salt should formally contain a Cu^{II} and a Cu^I ion, giving rise to a *mixed valence* complex. Cu^I was presumably obtained by the one-electron reduction of a Cu^{II} cation operated by the solvent. The crystal structure, shown in Figure 19.8,[168] indicates that the two copper centres are equivalent. Moreover, electron spin resonance (ESR) studies revealed that the unpaired electron is delocalised over the short Cu–Cu bond (2.36 Å), which discloses the presence of a σ bonding interaction between the two metal centres. Each copper possesses an average valence

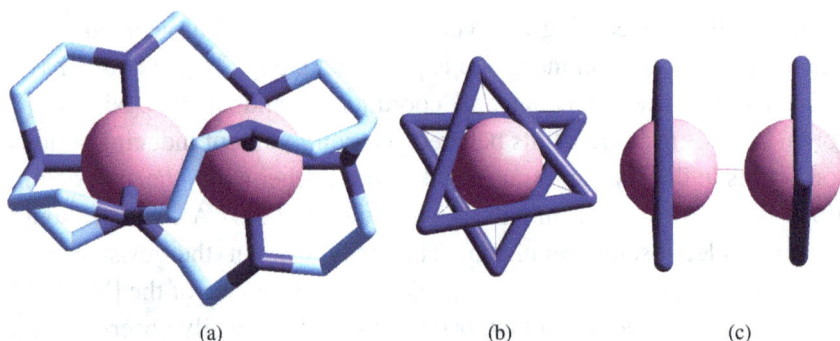

Figure 19.8. (a) Crystal structure of the $[Cu^{1.5}{}_2(bistren-C_2)](NO_3)_3 \cdot 2H_2O$ complex salt[168]; hydrogen atoms, nitrate counterions and solvating water molecules have been omitted for clarity; (b) triangles obtained by linking the nitrogen atoms of the secondary amine groups of each tren subunit; view along the ternary axis; (c) lateral view: the two $Cu^{1.5}$ ions protrude from the trigonal planes in order to establish an intermetallic σ bonding interaction; each $Cu^{1.5}$ ion is displaced from the trigonal plane by 0.14 Å.

of 1.5 and experiences a trigonal bipyramidal five-coordination, the donor set being constituted by the four nitrogen atoms of tren and by the other copper centre. Each $Cu^{1.5}$ protrudes by 0.14 Å from the equatorial N3 triangle of the bipyramid towards the other $Cu^{1.5}$, in order to establish a metal–metal covalent interaction. The average valence complex resists the oxidation to the dicopper(II) cryptate, a species probably destabilised by the Cu^{II}–Cu^{II} electrostatic repulsion.

The $[Cu^{1.5}{}_2(bistren-C_2)]^{3+}$ complex is indefinitely stable in water, but in MeCN disproportionates according to the following equation[163]:

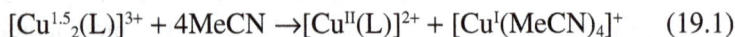

$$[Cu^{1.5}{}_2(L)]^{3+} + 4MeCN \rightarrow [Cu^{II}(L)]^{2+} + [Cu^{I}(MeCN)_4]^+ \qquad (19.1)$$

Disproportionation is favoured by the high stability of the tetrahedral complex $[Cu^{I}(MeCN)_4]^+$.

20

Size Exclusion Selectivity: The Case of Fluoride

The most valuable type of anion recognition is that related to size exclusion selectivity. Such a selectivity is observed when the receptor, providing for instance a spheroidal cavity, includes only spherical anions of radius less than or equal to a definite value. In this context, the smallest anion, fluoride, has offered a vast array of resources for exercise and investigation.

One of the first investigated receptors for selective fluoride encapsulation was the hexaprotonated form of the smallest bistren cryptand, **47** (bistren-C$_2$, L in the following). The interaction of F$^-$ with the polyammonium receptor was investigated by Lehn and coworkers through pH-metric titrations.[169] In the first experiment, a solution of L, containing an excess of a strong acid, was titrated with a standard base. In a subsequent experiment, the solution to be titrated also contained an excess of [Me$_4$N]F. The presence of fluoride caused a drastic change of the titration profile (pH vs mL of added base). Through nonlinear least-squares treatment of titration data, a log K value of 10.6 was calculated for the association equilibrium: LH$_6^{6+}$ + F$^-$ $\leftrightarrows$ [LH$_6\cdots$F]$^{5+}$. Such an extremely high stability in solution can be explained by looking at the crystal structure of the complex, shown in Figure 20.1.[170]

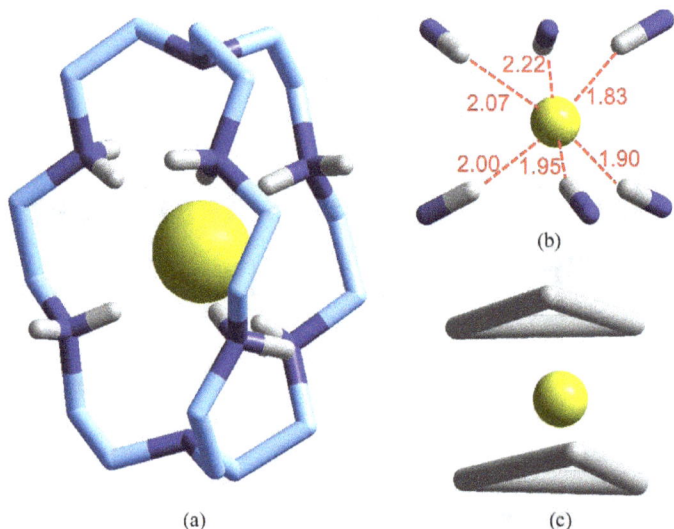

Figure 20.1. (a) Crystal structure of the complex salt $[LH_6\cdots F]Br_5\cdot 3H_2O$ (L = **47**, bistren-C_2);[170] C–H hydrogen atoms, bromide counterions and solvating water molecules have been omitted for clarity; (b) N–H fragments belonging to the secondary ammonium groups of LH_6^{6+} that point toward F$^-$; N–H$\cdots$F distances (in Å) correspond to hydrogen bonding interactions of moderate strength; (c) triangles have been obtained by linking the hydrogen atoms of the secondary ammonium groups of each tren subunit, involved in hydrogen bonding; the twist angle $\theta = 5° \pm 3°$ indicates a nearly regular trigonal prismatic coordination.

The fluoride ion receives six hydrogen bonds from the six secondary ammonium groups of LH_6^{6+}, all pointing towards fluoride. N–H$\cdots$F distances range over the 1.93–2.22 Å interval (moderate strength H-bonds). The triangles obtained by linking the anion-bound hydrogen atoms of each tren subunit form a twist angle $\theta = 5° \pm 3°$, which corresponds to an almost regular trigonal prismatic coordination.

Similar pH-metric titration experiments were also carried out for chloride. However, the presence of chloride excess did not significantly modify the titration curve for LH_6^{6+} alone, which allowed to estimate for the inclusion equilibrium a log $K \leq 2$.

Thus, LH_6^{6+} emerged as an excellent receptor for fluoride recognition in water. The extremely high F^-/Cl^- selectivity seemed to be ascribed to the different ionic radii of the two halides ($r_{F^-} = 1.33$ Å, $r_{Cl^-} = 1.81$ Å), the chloride being too big for easy inclusion into the hexaprotonated cryptand.

A few years later, Bowman-James and coworkers, by adding dropwise concentrated HCl to an MeOH solution of bistren-C_2, obtained a white precipitate, which, on recrystallisation in EtOH at 0°C, gave colourless prisms.[171] The corresponding crystal structure is shown in Figure 20.2.

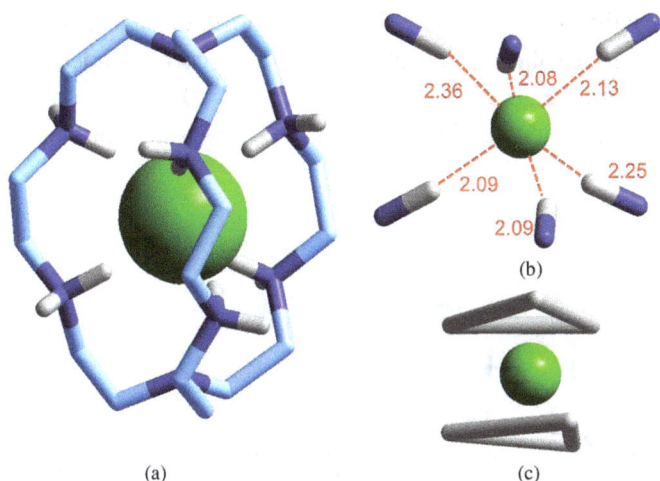

Figure 20.2. (a) Crystal structure of the complex salt $[LH_6 \cdots Cl]Cl_5 \cdot H_2O$ (L = **47**, bistren-C_2);[171] C–H hydrogen atoms, chloride counterions and solvating water molecule have been omitted for clarity; (b) N–H fragments belonging to the secondary ammonium groups of LH_6^{6+} that point toward Cl^-; N–H$\cdots$Cl distances (red dashed lines) in Å; (c) triangles have been obtained by linking the hydrogen atoms of the secondary ammonium groups of each tren subunit involved in hydrogen bonding; the twist angle $\theta = 24° \pm 4°$ is intermediate between trigonal prismatic ($\theta = 0°$) and octahedral ($\theta = 60°$) coordination.

Chloride is well encapsulated in the receptor's cavity. It receives six H-bonds from the six secondary ammonium groups, exactly like F^- (Figure 20.1), while N–H$\cdots$Cl distances are those expected for the formation of H-bonds of moderate strength. The coordination geometry is not exactly the same as for fluoride, but it is at midway between the trigonal prism and the octahedron ($\theta = 24° \pm 4°$). Thus, the cryptand framework, probably due to the presence of the $-CH_2CH_2-$ spacers, seems flexible enough to accommodate anions of distinctly different size and, even if highly selective in favour of fluoride, it does not exert size exclusion selectivity. The much higher stability of the fluoride complex mainly results from the higher basicity of F^- with respect to Cl^- and from the consequent capability of F^- to establish stronger hydrogen bonding interactions. Bromide ($r_{Br^-} = 1.96$ Å) is probably too large to be encapsulated by LH_6^{6+}. In the isolated crystalline complex salt $[LH_6]Br_6 \cdot H_2O$, the cavity is empty and the six bromide ions lie outside.[170]

On reaction of bistren-C_2 with triethylorthoformate at 120°C in dry xylene, a white solid can be obtained, which corresponded to the tris-imidazolidinium cage **51**, L^{3+}, shown in Figure 20.3.[172] L^{3+} contains three imidazolidinium subunits integrated in the bistren skeleton.

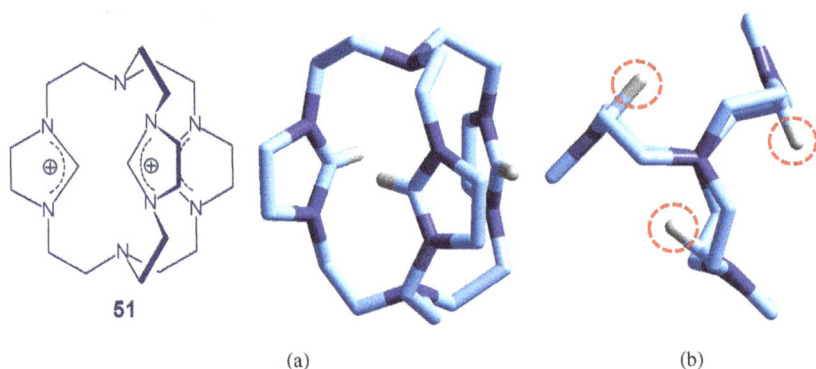

51

(a)   (b)

Figure 20.3. The crystal structure of $[L](ClO_4)_3 \cdot H_2O$ ($L^{3+} = $ **51**):[172] (a) lateral view; (b) top view. Only the hydrogen atoms of the imidazolinium subunits are shown; perchlorate counterions and solvate water molecule have been omitted for clarity.

The C–H fragment of imidazolidinium is highly polarised and can behave as an effective H-bond donor for anions. However, the crystal structure of the $[L](ClO_4)_3 \cdot H_2O$ salt,[172] shown in Figure 20.3, indicates that the three C–H fragments do not point towards the cavity, but outside, a circumstance unfavourable to anion encapsulation. The interaction of L^{3+} with fluoride was investigated through ^1H NMR titration with NaF of a

D_2O solution of the tris-imidazolidinium receptor **51**, made strongly acidic (pD = 1). The titration profile disclosed the formation of a 1:1 complex, but the profile was too steep to allow a safe determination of the association constant. Thus, a competitive titration was carried out under the same conditions in the presence of Zr^{IV} (competing for F^- through the formation of the $[Zr^{IV}F]^+$ complex) and a log K = 12.5 was determined for the complex formation equilibrium. Moreover, from a solution of **51**, made acidic with $HClO_4$, containing BF_4^-, which slowly released F^- to the solution, the complex salt $[LH_2\cdots F](ClO_4)_2(BF_4)_2 \cdot H_2O$ crystallised (L^{3+} = **51**). The corresponding structure is shown in Figure 20.4.[172]

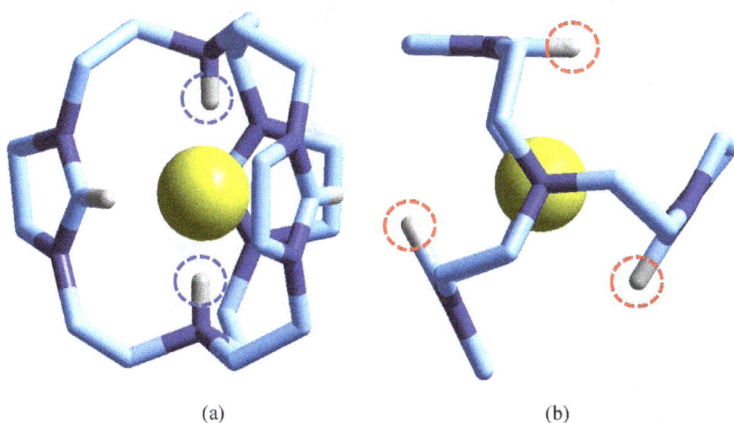

(a)   (b)

Figure 20.4. The crystal structure of $[LH_2\cdots F](BF_4)_2(ClO_4)_2 \cdot H_2O$ (LH_2^{5+} = diprotonated form of **51**)[172]: (a) lateral view; (b) top view. Only the acidic hydrogen atoms of the imidazolinium subunits (three C–H, two N–H) are shown; perchlorate and tetrafluoborate counterions and solvate water molecule have been omitted for clarity.

The receptor has a 5+ charge, due to the protonation of the two tertiary amine nitrogen atoms, (LH_2^{5+}). The fluoride ion is well incorporated in the LH_2^{5+} cavity, but it does not interact with the imidazolidinium N–H fragments, which still point outside of the cavity, as observed for the uncomplexed receptor L^{3+}. F^- receives two H-bonds from the protonated tertiary amine nitrogen atoms, both exhibiting an *in* configuration. The N–H$\cdots$F distances (1.68 and 1.74 Å) are remarkably shorter than those observed in the fluoride complex of the hexaprotonated bistren-C_2, which range over the 1.93–2.22 Å interval, and are presumably stronger. Fluoride binding does not seem to induce any serious rearrangement of the receptor, as judged from the distance between the two pivot nitrogen atoms, 5.23 Å, only slightly smaller than that observed in the uncomplexed unprotonated system L^{3+} (5.43 Å). Very interestingly, titration with Cl^- under the same conditions (pD = 1) did not induce any modification in the ^1H NMR spectrum of the receptor. Thus, LH_2^{5+} exerts specific recognition of F^-. Notice also that LH_2^{5+} rightfully belongs to the ancient class of katapinands. The imidazolidinium subunits do not directly contribute to the anion binding with H-bonds, but undoubtedly increase the intensity of the electrostatic interactions. However, it is probably their rigidity that affords size exclusion selectivity in favour of the smallest anion.

21

Pseudo-Cryptands: Closing the Receptor's Cavity with a Transition Metal

A tripodal ligand L (e.g. tren) can be converted into a cryptand by appending at each of its three arms a molecular fragment capable of behaving as a ligand towards a given metal ion M. Binding of M by these ligands generates a closed space which can host a suitable substrate. The process is pictorially illustrated in Figure 21.1. In the first reported example, three 2,2′-bipyridine fragments (bpy) were appended to the secondary amine nitrogen atoms of tren via a $-CH_2-$ spacer in position 5, to give **52** (L).[173] Fe^{II} was chosen as a 'closing' metal, as it forms a very stable $[Fe^{II}(bpy)_3]^{2+}$ complex, of octahedral geometry, $3d^6$ low-spin. Stability is derived from the σ donation from pyridine type nitrogen atoms to the metal and from π back-donation from filled $d\pi$ orbitals of Fe^{II} to low energy empty π antibonding molecular orbitals delocalised over the bpy framework. In addition, the $[Fe^{II}(bpy)_3]^{2+}$ complex is kinetically inert, which guarantees the function of iron(II) as a 'glueing' metal. This study was conducted to verify whether the $[Fe^{II}(L)]^{2+}$ complex (L = **52**) could be

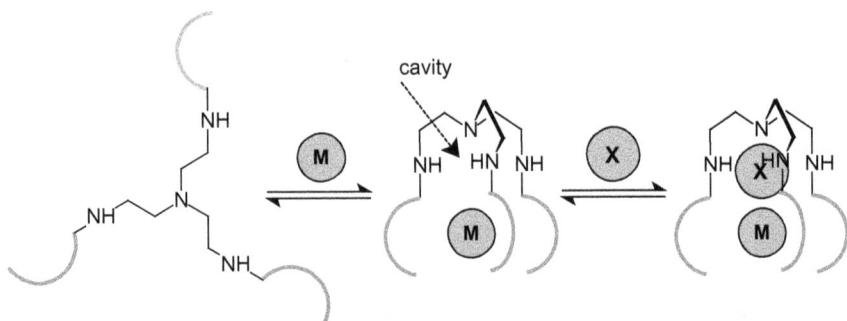

Figure 21.1. The synthesis of a pseudo-cryptand and of its complex. Ligating fragments eager to coordinate a given metal ion M (half circles) are covalently linked to the arms of tren to give L. On formation of the ML complex, room is available inside the tren subunit for the inclusion of any substrate X.

able to host in its tren cavity another transition metal, e.g. Cu^{II}. The stepwise process is pictorially represented in Figure 21.2.

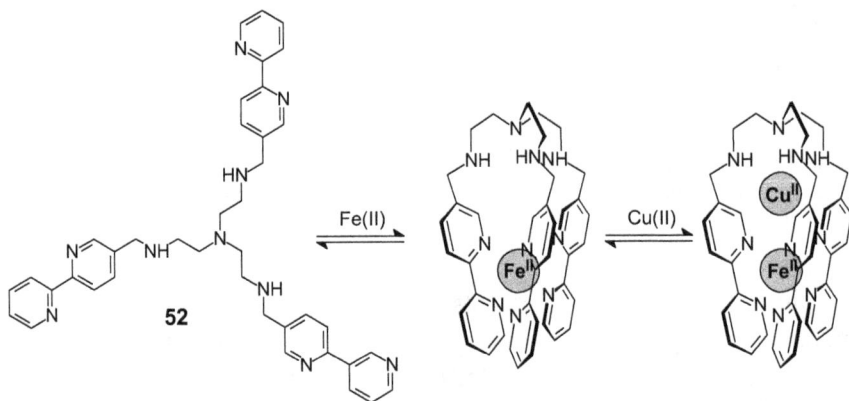

Figure 21.2. The synthesis of a tren-bpy pseudo-cryptand suitable for the inclusion of Cu^{II} in the space subtended by the tren subunit.

On mixing in an aqueous solution of stoichiometric amounts of **52** (L) and $Fe^{II}SO_4$, the solution took an intense purple-red colour. On addition of NH_4PF_6, a purple complex salt of formula $[Fe^{II}(LH)(PF_6)_3$ precipitated. Its crystal structure is shown in Figure 21.3.

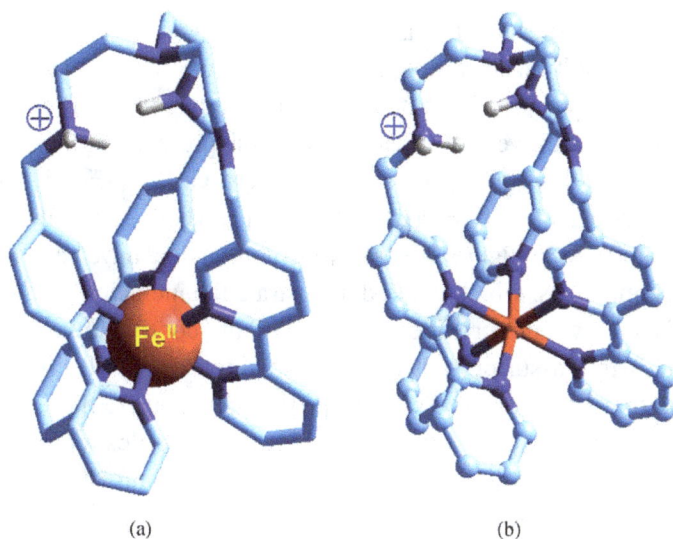

(a)   (b)

Figure 21.3. (a) The crystal structure of the complex salt $[Fe^{II}(LH)](PF_6)_3 \cdot MeCN$ (L = **52**).[173] C–H hydrogen atoms, hexafluorophosphate counterions, and solvate MeCN molecule have been omitted for clarity. Ball and stick rendering in (b) makes evident the octahedral geometry of the $Fe^{II}(bpy)_3^{2+}$ subunit.

One of the secondary amine groups of the tren moiety is protonated, while the $Fe^{II}(bpy)_3^{2+}$ subunit shows octahedral coordination geometry. The structure shows that the conformational arrangement of the pseudo-cryptand leaves room for the inclusion of a further metal ion. $[Fe^{II}(L)]^{2+}$ maintains the properties of the parent complex $Fe^{II}(bpy)_3^{2+}$. In particular, it is indefinitely stable in 0.1 M $HClO_4$ and in 0.1 M NaOH, thus granting the permanent integrity of the pseudo-cryptand cavity.

pH-metric titration of an acidic solution of $[Fe^{II}(L)]^{2+}$ with standard base showed that the tren subunit behaves as a triprotic base with a log K_i sequence of stepwise protonation equilibria: 8.90, 7.48 and 5.69. Each value is distinctly lower than the corresponding one observed for the stepwise protonation of tren (10.15, 9.45 and 8.43), due to the intramolecular electrostatic repulsions of the ammonium groups in a constrained system between themselves and with the proximate Fe^{II} ion.

A titration experiment was then carried out on a similar solution containing in addition an equimolar amount of Cu^{2+}. Substantial modification of the titration curve indicated the formation of copper(II) complexes. Best fitting of the curve was consistent with the presence of the following copper(II) containing species: $[Fe^{II}(LH)\cdots Cu]^{5+}$, $[Fe^{II}(L)\cdots Cu]^{4+}$, $[Fe^{II}(L)\cdots Cu(OH)]^{3+}$, and $[Fe^{II}(L)\cdots Cu(OH)_2]^{2+}$, whose concentration profiles are displayed in Figure 21.4(a). No structural data have been reported for these complexes, which could demonstrate the inclusion of Cu^{II} in the receptor's cavity and provide information on coordination geometry. However, it is reassuring that copper(II) complexes of $[Fe^{II}(L)]^{2+}$ form according to a stoichiometry and a sequence analogous to those observed for bistren cryptands, e.g. with **28**. Concentration profiles referring to the titration of an acidic solution containing bistren cryptand **28** and Cu^{2+} in a 1:1 molar ratio are shown in Figure 21.4(b).

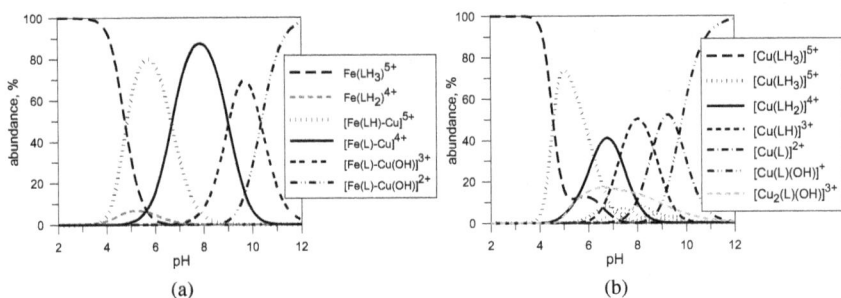

Figure 21.4. Percent concentrations of the species present at the equilibrium over the course of the titration with standard base of a solution containing equimolecular amounts of the receptor and Cu^{2+}, plus excess acid: (a) receptor = $[Fe^{II}(L)]^{2+}$ (L = **52**); (b) receptor = L, (**28**).

The dimetallic complex $[Fe^{II}(L)\cdots Cu]^{4+}$ is prevalent in the 6.6 < pH < 8.9 range, reaching 90% at pH 8. On comparing its formation constant (log $K = 9.06$) with that found for the equilibrium $Cu^{2+} + $ tren $\leftrightarrows [Cu^{II}(tren)]^{2+}$ (log $K = 18.50$), a dramatic decrease is observed. The lowered stability has to be ascribed to the electrostatic repulsion exerted towards Cu^{2+} by the proximate iron(II), only partially shielded by bpy molecules. However, a more exact comparison should be made with the equilibrium $[Cu(L)]^{2+} + $

$Cu^{2+} \leftrightharpoons [Cu^{II}_2(L)]^{2+}$ (L = **28**), in which the second Cu^{II} ion enters the neutral tren subunit of the monometallic $[Cu(L)]^{2+}$ cryptate complex: log $K =$ 9.41 (see structural sketches in Figure 21.5).

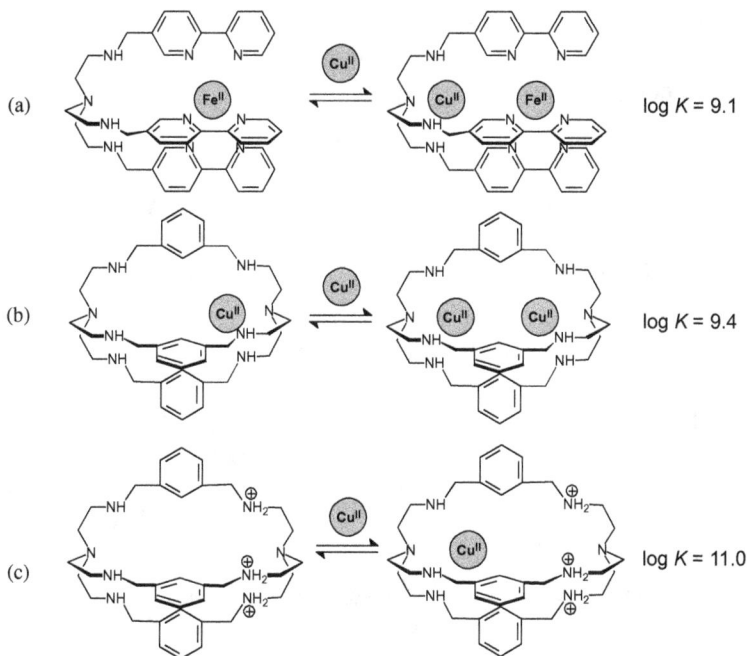

Figure 21.5. Entering of a Cu^{2+} ion into a pseudo-cryptand or cryptand to occupy and empty tren cavity, facing the electrostatic repulsions by: (a) an $Fe^{II}(bpy)_2{}^{2+}$ moiety; (b) another Cu^{II} ion; (c) a triprotonated tren subunit.

The closeness of log K values suggest that $Fe^{II}(bpy)_2{}^{2+}$ and $Cu^{II}(tren)^{2+}$ moieties exert electrostatic repulsions of approximately the same magnitude. Process (c) in Figure 21.5 refers to the entering of Cu^{2+} into cryptand **28** in its triprotonated form: in this case, the metal faces the three positive charges of the three secondary ammonium groups, but their repulsive effects are apparently reduced by the greater distance. Thus, the inclusion process results favoured by two orders of magnitude with respect to equilibria in Figures 21.5(a) and 21.5(b).

No further studies have appeared on a metal that traps another metal, an intrinsically unnatural phenomenon. More attention has been later devoted to the design of metal-locked cryptands suitable for anion inclusion. Quite comfortably, in the present case, the cation/anion electrostatic attraction is expected to favour the formation of the complex. Along this line, Nicholas Fletcher, Queen's University, Belfast, reported the synthesis of tripodand **53** (L), a tren amide with three appended bpy, which was locked with a ruthenium(II) ion.[174] The structure of the $[Ru^{II}(L)]^{2+}$ pseudocryptate complex is shown in Figure 21.6.

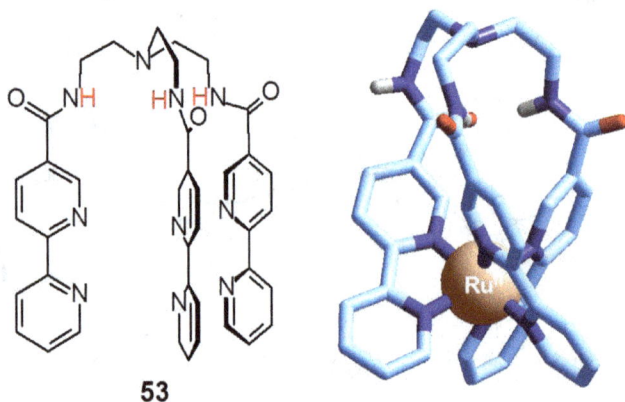

53

Figure 21.6. The crystal structure of the complex salt $[Ru^{II}(L)](PF_6)_2$ (L = **53**).[174] C–H hydrogen atoms and hexafluorophosphate counterions have been omitted for clarity.

Ru^{II}, $4d^6$ low-spin, forms with bpy an octahedral complex that is very stable from both a thermodynamic and kinetic point of view. $Ru^{II}(bpy)_3^{2+}$ is the most classical luminescent complex. It shows an intense red-orange colour (absorption band centred at 450 nm, molar extinction coefficient $\varepsilon = 15000$ M^{-1} cm^{-1}); when excited at 450 nm, it emits a yellow-orange luminescence (emission band centred at 627 nm), it exhibits a lifetime $\tau = 627$ ns and a quantum yield $\Phi = 0.042$ (of 100 absorbed photons, 4 are emitted — compare with the classical organic fluorophore anthracene, excitation at 355 nm, emission at 405 nm, blue fluorescence, $\tau = 5$ ns, $\Phi = 0.36$). The excited electron in $*Ru^{II}(bpy)_3^{2+}$ lies on a π^* orbital of a bpy molecule, which makes it available to fast electron Transfer and Energy Transfer processes with a variety of reactants through an outer sphere mechanism.[175]

The $Ru^{II}(bpy)_3^{2+}$ subunit in $[Ru^{II}(L)]^{2+}$ (L = **53**) showed a slightly distorted octahedral coordination geometry and displayed the typical spectral properties (λ_{em} = 634 nm, Φ = 0.029, in MeCN at 25°C). The tren amide cavity in the upper floor of the complex, possessing three polarised N–H fragments, appeared potentially suitable for anion inclusion. However, addition of Cl^-, Br^- and NO_3^- did not cause any significant modification of the emission spectrum of $[Ru^{II}(L)]^{2+}$ in MeCN, thus excluding anion inclusion in the tren amide cavity.

At this stage, one could ask why the amine tren cavity of the Fe^{II}-locked pseudocryptand **52** includes Cu^{II}, whereas the amide tren cavity of the Ru^{II}-locked pseudocryptand **53** does not include Cl^-. The two pseudocryptands are expected to have cavities of comparable volumes. In fact, crystal structures say that the three amide nitrogen atoms of the Ru^{II} derivative lie on the corners of a nearly regular equilateral triangle with edges of length 4.4 ± 0.2 Å, while the triangle made by the three amine nitrogen atoms of the Fe^{II} pseudocryptand has an average edge of 4.0 ± 0.1 Å. Thus, the inclusion of a substrate, if anything, should be easier for the amide cavity than for the amine cavity. The quite obvious answer is that the different behaviour depends upon the size of the envisaged substrate: Cu^{II}, ionic radius for five-coordination 0.79 Å (0.71 Å for four-coordination),[176] is suitable for incorporation by the amine cavity; Cl^-, ionic radius 1.81 Å, is too large for inclusion cavity. These considerations prompted Fletcher and coworkers to design a receptor having as a capping subunit an 1,3,5-aminomethylbenzene subunit, whose structural formula is reported in Figure 21.7 (**54**).[174]

The receptor offers six secondary amide groups for anion coordination, while the 1,3,5-substituted benzene cap is expected to provide larger room for anion encapsulation than the pseudo-cryptate derived from **53**. Indeed, the crystal structure of the ruthenium(II) complex salt $[Ru^{II}(L)]$ $Br_2 \cdot 3H_2O$ (L = **54**) in Figure 21.7(a) shows an opening of size suitable for the inclusion of mono- and polyatomic anions. In particular, the three nitrogen atoms of the 'lower' amide groups (those more distant from the cap) lie on the corners of a triangle of edges: 9.15, 8.92, and 8.00 Å.

^{1}H NMR titration were carried out in CD_3CN that showed that NO_3^- (log K = 6.4) has a much greater affinity for $[Ru^{II}(L)]^{2+}$ (L = **54**) than

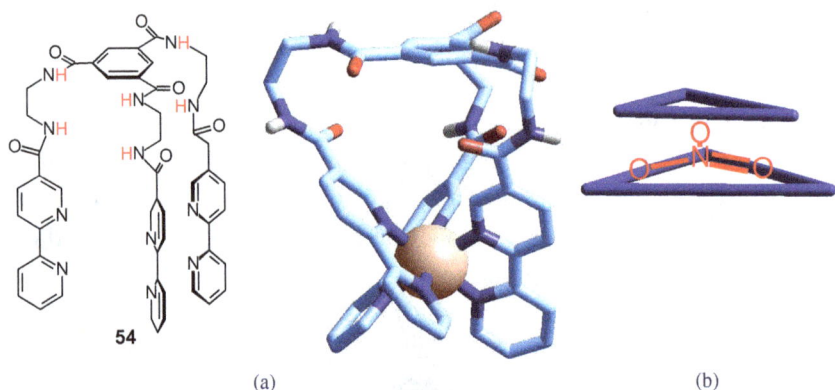

54

(a)   (b)

Figure 21.7. (a) The crystal structure of the complex salt $[Ru^{II}(L)]Br_2 \cdot 3H_2O$ (L = **54**).[174] C–H hydrogen atoms, bromide counterions and solvate water molecules have been omitted for clarity. (b) Triangles obtained by linking the three upper amide hydrogens and the three lower amide hydrogens. The nitrate ion can place itself in the space between the two triangles in a sandwich mode in order to profit from H-bond interactions.

Cl^- (log K = 2.1) and Br^- (log K = 2.9). It can be suggested that the triangular nitrate ion places itself in the narrow space between the two triangles defined by the amide nitrogen atoms, in a *sandwich* mode, as tentatively illustrated in Figure 21.7(b), thus profiting from multiple hydrogen bonding interactions. Chloride and bromide cannot insinuate themselves in the narrow space between the two assemblies of amide groups, which may explain the lower values of log K.

The red-orange MeCN solution of complex $[Ru^{II}(L)]^{2+}$ (L = **54**) in MeCN emits a yellow-orange luminescence at 640 nm, with a quantum yield Φ = 0.038. On anion addition, luminescent emission was partially and selectively quenched. In particular, on addition of Cl^- (10 equivalent), the emission intensity was reduced to 70%, and for Br^- to 80%. Such a behaviour does not necessarily imply that the halide ions are included in the cavity. They may quench the excited state even if associated with the pseudocryptand as an ion pair. In this regard, it should be considered that in the crystalline complex $[Ru^{II}(L)]Br_2 \cdot 3H_2O$, whose structure is illustrated in Figure 21.7(a), the bromide ion is not included in the pseudocryptand, but is placed outside in the lattice.

Quite surprisingly, NO_3^-, even if well incorporated in the pseudocryptand, as the high log K value suggests, did not induce any significant modification of the emission spectrum of an MeCN solution of $[Ru^{II}(L)]^{2+}$. This may be due to the fact that nitrate, trapped by the amide network, is kept far away from the $Ru^{II}(bpy)_3^{2+}$ subunit and that, in any case, its orientation does not favour the proper orbital overlap for the occurrence of an electron or energy transfer process.

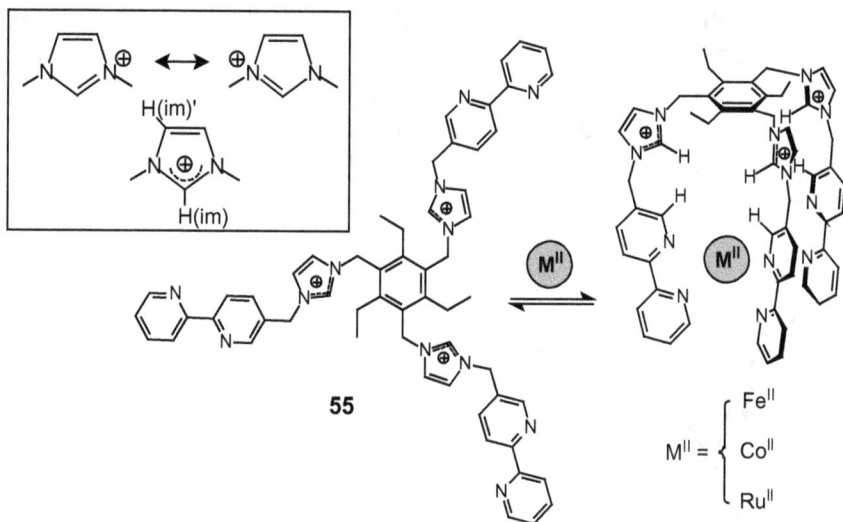

Figure 21.8. The synthesis of the tripodal ligand **55**, containing a 1,3,5-trimethylbenzene cap. Reaction of **55** with Fe^{II}, Co^{II} or Ru^{II} gives a metal-locked pseudocryptand.[177] Six polarised C–H fragments (three from imidazolium subunits (H(im)), three from pyridine rings (H(py)) are available for hydrogen bonding interactions with an included anion.

A complete investigation on the behaviour of metal-locked pseudocryptands with respect to anion inclusion and recognition was based on tripodand **55**, shown in Figure 21.8, which consists of a 1,3,5-trimethylbenzene cap, to which three arms have been appended, each constituted by an imidazolium subunit linked to a 2,2′-bipyridine fragment.[177] Coordination of the three bpy fragments to a given metal ion was expected to generate an upper cavity suitable for anion inclusion. In fact, six polarised C–H fragments — three from the imidazolium subunits, H(im), and three from the closer pyridine rings, H(py), polarised through coordination of the pyridine ring to M^{II} — point towards the middle of the cavity, ready to interact with a hypothetical included anion.

^{1}H NMR titration experiments in a variety of media disclosed a definite tendency of the $[Fe^{II}(L)]^{5+}$ receptor (L = **55**) to include halide ions. In particular, on titration with $[Bu_4N]Br$ of a solution 10^{-3} M of $[Fe^{II}(L)](PF_6)_5$ in a CD_3CN/D_2O mixture (4:1 v/v), a sharp downfield shift of the C–H(py) protons was observed, indicating direct interaction with the anion. The C–H(py) proton owes its H-bond donor tendency to the polarisation exerted by the bpy-bound metal centre. The signal pertinent to the C–H(im) protons was not observed, probably due to the occurrence of a fast exchange with D_2O. However, involvement of the C–H(im) fragments in anion binding was indirectly suggested by the moderate, yet well distinguishable upfield shift of the C–H(im)′ proton of the imidazolium ring (see structural formula in Figure 21.8).

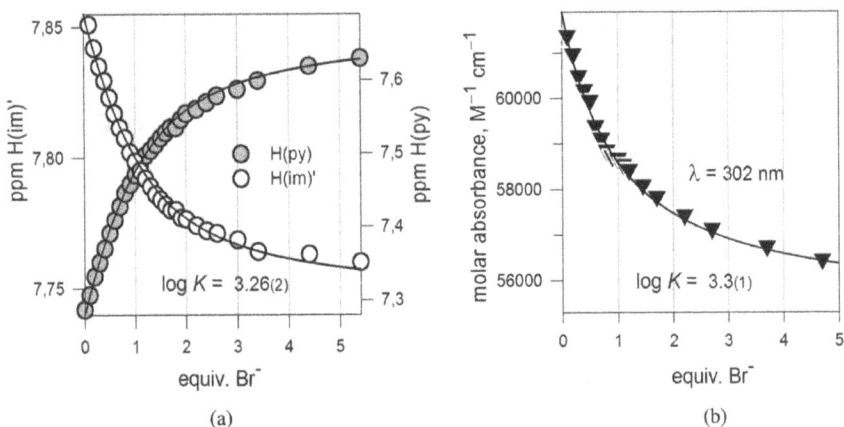

Figure 21.9. Titrations of an MeCN/H_2O 4:1 (v/v) solution of $[Fe^{II}(L)](PF_6)_5$ (L = **55**) with $[Bu_4N]Br$: (a) ^{1}H NMR titration of a solution 10^{-3} M; (b) UV titration of a solution 10^{-4} M.[177]

Figure 21.9(a) shows the corresponding titration profiles based on the chemical shifts C–H(py) protons and of C–H(im)′ protons, from which a log K = 3.26 ± 0.02 was calculated for the association equilibrium: $[Fe^{II}(L)]^{5+} + Br^- \leftrightarrows [Fe^{II}(L)(Br)]^{4+}$.

The interaction of $[Fe^{II}(L)]^{5+}$ with Br^- was also monitored through a moderate yet significant modification of the UV spectrum. In particular, on bromide addition to a solution 10^{-4} M in $[Fe^{II}(L)](PF_6)_5$, the intensity of the band at 302 nm, originating from a π–π* transition centred on pyridine rings, was observed to decrease following the hydrogen bonding interaction of C–H(py) with Br⁻. The corresponding titration profile is shown in Figure 21.9(b), from which log $K = 3.3 \pm 0.1$ was calculated.

Unambiguous information on the nature of the receptor–anion complex came from X-ray diffraction studies on a single crystal of the red-violet $[Fe^{II}(L) \cdots Br](PF_6)_4 \cdot 2H_2O \cdot MeCN$ salt. The structure of the bromide inclusion complex is shown in Figure 21.10(a).[177]

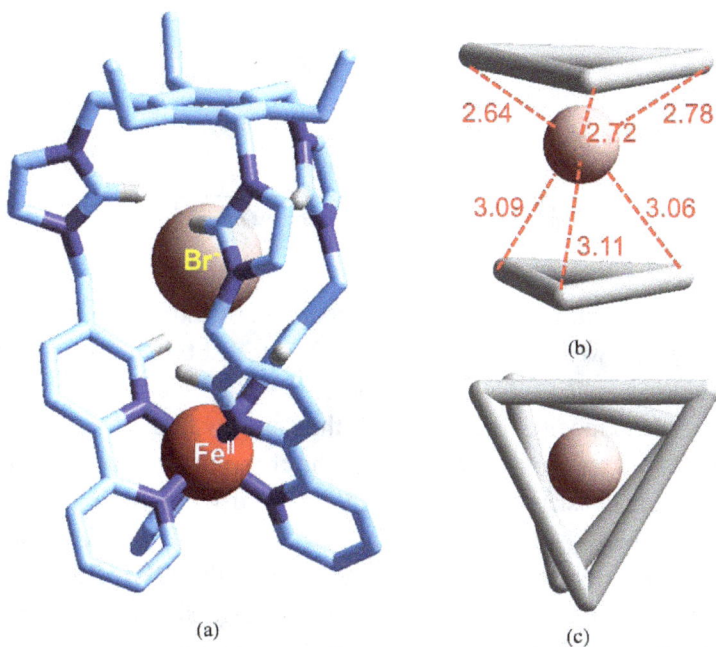

Figure 21.10. (a) The crystal structure of $[Fe^{II}(L) \cdots Br](PF_6)_4 \cdot MeCN \cdot 3H_2O$ (L = **55**)[177]; all hydrogens but H(im) and H(py) atoms, hexafluorophosphate counterions and solvate molecules have been omitted for clarity; (b) nearly equilateral triangles obtained by linking the three H(im) (upper) and the three H(py) (lower), with H⋯Br distance in Å; (c) top view.

The bromide ion is included within the tris-imidazolium cavity and is placed on the line that connects the centroid of the 1,3,5-triethylbenzene platform and the Fe^{II} ion. It receives three hydrogen bonds from the C–H(im) fragments of the three imidazolium subunits (average H⋯Br distance 2.71 ± 0.07 Å) and three from the C–H(py) fragments of the close pyridine rings (average H$\cdots$Br distance 3.07 ± 0.04 Å). The six coordinated hydrogen atoms are positioned at the corners of a slightly distorted trigonal prism (twist angle $\theta = 12.4°$).

Table 21.1. Log K values for the equilibria: $[Fe^{II}(L)]^{5+} + X^{-1}$ $\leftrightarrows [Fe^{II}(L)(X)]^{4+}$ (L = **55**).

X^-	MeCN/H$_2$O 4:1	MeCN
Cl$^-$	3.80(3) NMR, 3.9(1) UV-vis	>7 UV-vis
Br$^-$	3.26(1) NMR, 3.3(1) UV-vis	>7 UV-vis
I$^-$	—	4.40(3) UV-vis
NCO$^-$	4.2(1) UV-vis	
NCS$^-$	4.09(2) UV-vis	
N$_3^-$	5.7(1) UV-vis; >5 NMR	

Note: In parentheses is the uncertainty on the last figure.[177]

Titration experiments were extended to other halides. Log K for Cl$^-$ in MeCN/H$_2$O 4:1 was 3.80 ± 0.03 (^{1}H NMR) and 3.9 ± 0.1 (UV-vis). Addition of iodide to an aqueous MeCN solution of $[Fe^{II}(L)]^{5+}$ did not cause any significant spectral modification. However, the same titration carried out in pure MeCN induced a significant decrease of the absorption band at 300 nm and an equilibrium constant log $K = 4.40 \pm 0.03$ was determined (in pure MeCN Cl$^-$ and Br$^-$ gave log $K > 7$). Log K values for $[Fe^{II}(L)]^{5+} + X^- \leftrightarrows [Fe^{II}(L)\cdots X]^{4+}$ equilibria in different media are displayed in Table 21.1. The observed sequence of affinity (Cl$^-$ > Br$^-$ > I$^-$) reflects the decreasing tendency to establish H-bonds and, ultimately, the decrease of anion basicity.

The network of six polarised C–H fragments of $[Fe^{II}(L)]^{5+}$ seems ideal for the incorporation of spherical anions (halides) and much less favourable to the inclusion of polyatomic anions. Nevertheless, a well-defined interaction pattern was observed upon titration of an MeCN solution of $[Fe^{II}(L)]^{5+}$ with the polyatomic rod-like pseudohalides. In particular, on progressive addition NaN$_3$ to a CD$_3$OD solution of $[Fe^{II}(L)](PF_6)_5$, a distinct downfield shift of the C–H(py) signal was observed. In particular, the titration profile indicated the formation of a 1:1 complex, but the absence of curvature prevented the determination of the association constant.

A value of log $K = 5.7 \pm 0.1$ was determined through a spectrophotometric titration experiment carried out on a solution MeCN/H$_2$O 4:1 v/v 10^{-4} M in [Fe^{II}(L)](PF$_6$)$_5$, by monitoring the decrease in the intensity of the π–π* band centred at 300 nm. Noticeably, the association constant for N$_3^-$ is 80-fold higher than that of Cl$^-$ and also distinctly higher than for the other pseudohalides, NCO$^-$ and NCS$^-$ (log K values in Table 21.1).

Insights on the unexpectedly high stability of the [Fe^{II}(L)$\cdots$N$_3$]$^{4+}$ complex in solution are provided by the crystal structure of the salt [Fe^{II}(L)$\cdots$N$_3$](PF$_6$)$_4 \cdot$ 1.5H$_2$O $\cdot$ 3MeCN (L = **55**), shown in Figure 21.11.[177]

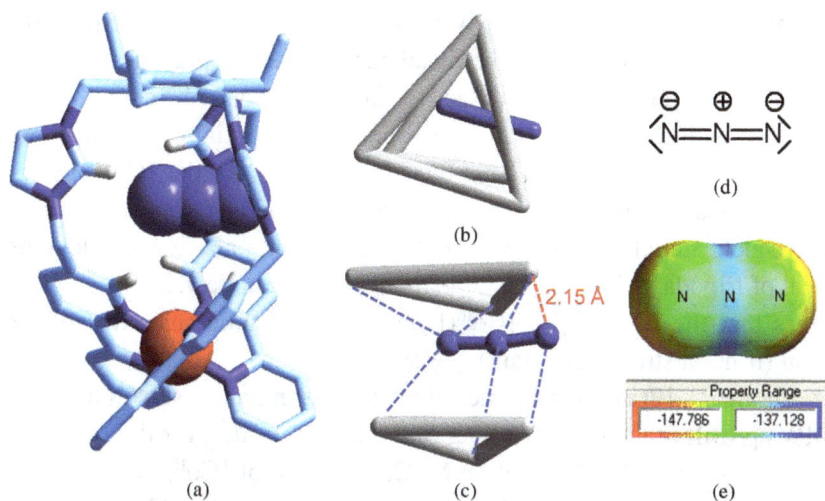

Figure 21.11. (a) The crystal structure of the complex salt [Fe^{II}(L)$\cdots$N$_3$] (PF$_6$)$_4 \cdot$ 1.5H$_2$O $\cdot$ 3MeCN (L = **55**);[177] all hydrogens but H(im) and H(py) atoms, hexafluorophosphate counterions and solvate molecules have been omitted for clarity; (b) nearly equilateral triangles obtained by linking the three H(im) (upper) and the three H(py) (lower); top view; (c) lateral view: dashed lines indicate hydrogen bonds, only one of moderate energy (red line, 2.15 Å); blue lines indicate hydrogen bonding interactions of low energy, to which C–H$\cdots$N distances ranging from 2.57 to 2.89 Å correspond; (d) the Lewis structural formula of N$_3^-$; (e) distribution of the electrostatic potential as obtained by *ab initio* calculations, colour scale is in kcal mol^{-1}.

The azide ion is not completely inserted in the pseudocryptand cavity, and one terminal nitrogen atom protrudes from the cavity. N$_3^-$ receives a total of seven H-bonds, one of moderate intensity (red dashed line in Figure 21.11(c), C–H$\cdots$N distance 2.15 Å), the others of low intensity, with C–H$\cdots$N distances ranging from 2.57 to 2.89 Å. The high stability of the [Fe^{II}(L)$\cdots$N$_3$]$^{4+}$ complex seems to be related to the inherent tendency of azide to establish multipoint hydrogen bonds. Noticeably, also the central nitrogen atom of N$_3^-$ participates in the hydrogen bonding network receiving two H-bonds. This contrasts with the typical Lewis representation (Figure 21.11(d)), in which the central nitrogen atom detains a formally positive charge. However *ab initio* calculations demonstrated that a negative electrostatic potential is distributed over the whole surface of the ion, even if the central nitrogen atom is distinctly less negative (see Figure 21.11(e)).

In the previously described example, the locking metal, Fe^{II}, played a purely architectural role. But the metal can be made to play an additional role. Thus, if the metal is redox active through the M^{II}/M^{III} couple and if the anion inclusion modifies the corresponding potential $E°$(M^{III}/M^{II}), we are in the presence of an *electrochemical sensor*. As the [Co^{II}(bpy)$_3$]$^{2+}$ complex undergoes a metal centred reversible one-electron oxidation to the trivalent state ($E° = 0.03$ V vs Fc$^+$/Fc), Co^{II} was chosen as the locking metal of tripodand **55**.[178] In fact, the moderately positive value of $E°$ guarantees that the [Co^{II}(L)]$^{5+}$→[Co^{III}(L)]$^{6+}$ oxidation anticipates that of the anion (if not a strong reductant).

Figure 21.12(a) shows some selected differential pulse voltammetry (DPV) profiles taken at a working platinum electrode over the course of the titration with [Bu$_4$N]Cl of an MeCN solution of [Co^{II}(L)]$^{5+}$ (L = **55**) (potential E vs the Fc$^+$/Fc reference couple). Before chloride addition, a current peak is observed at 100 mV vs Fc$^+$/Fc, E, pertaining to the Co^{II}→Co^{III} oxidation of the empty pseudocryptand On adding Cl$^-$, the intensity of the peak decreases, while a new peak develops at –0.44 mV. Such a peak originates from the Co^{II}→Co^{III} oxidation with chloride included in the cavity (E_X). On further anion addition, the peak at 100 mV disappears, while that at –0.44 mV increases to reach a limiting value (see the family of DPV profiles in Figure 21.12(a)). The diminishing of the peak potential reflects the reduction of the electrical work associated to the +2/+3 charge increase in the presence of a proximate negative charge. More precisely, the difference of the peak potentials. E_X – $E = \Delta E$, is related to the ratio of the association constants K^{III} (referring

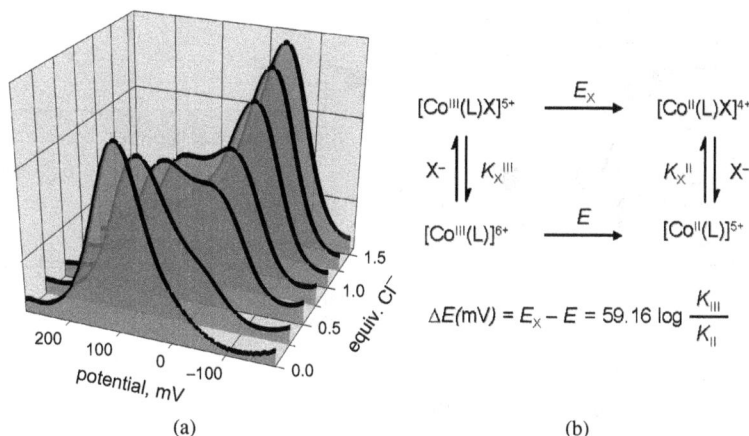

$$\Delta E(mV) = E_x - E = 59.16 \log \frac{K_{III}}{K_{II}}$$

(a)   (b)

Figure 21.12. (a) Family of DPV profiles taken at a platinum electrode over the course of the titration with [Bu$_4$N]Cl of an MeCN solution of [Co^{II}(L)]$^{5+}$ (L = **55**) (potential E (mV) vs the Fc$^+$/Fc reference couple), potential scanning from negative to positive potentials (10 mV s^{-1}); on chloride addition, the DPV peak at 100 mV, pertaining to the oxidation of [Co^{II}(L)]$^{5+}$ to [Co^{III}(L)]$^{6+}$ (empty cavity) decreases in intensity, while a new peak develops and grows at -0.44 mV, which results from the oxidation of the chloride inclusion complex: [Co^{II}(L$\cdots$Cl)]$^{4+} \rightarrow$ [Co^{III}(L)$\cdots$Cl]$^{5+}$ + e$^-$;[178] (b) square scheme illustrating the relationship between the peak separation ΔE (potential in the presence of the anion X$^-$ – potential in the absence of X$^-$) and the anion association constants of the oxidised (K^{III}) and of the reduced (K^{II}) receptors.

to the [Co^{III}(L)]$^{6+}$ receptor) and K^{II} (referring to the ([Co^{II}(L)]$^{5+}$ receptor), as illustrated by the square diagram in Figure 21.12(b). In particular, to a $\Delta E = 144$ mV, a value of log $(K^{III}/K^{II}) = 2.44$ corresponds (value calculated from the equation in Figure 21.12(b)). The fact that K^{III} is 275-fold greater than K^{II} is due to the greater metal–anion electrostatic attraction, but also to the increase of the H-bond donor properties of the C–H(py) fragments, which have been more intensely activated by the proximate low-spin Co^{III} centre.

Bromide displays an analogous behaviour. Figure 21.13 shows the DPV profiles obtained at a platinum microsphere working electrode over the course of the titration with [Bu$_4$N]Br of an MeCN solution of [Co^{II}(L)]$^{5+}$ (L = **55**) (potential E (mV) vs the Fc$^+$/Fc reference couple), potential scanning from negative to positive potentials.

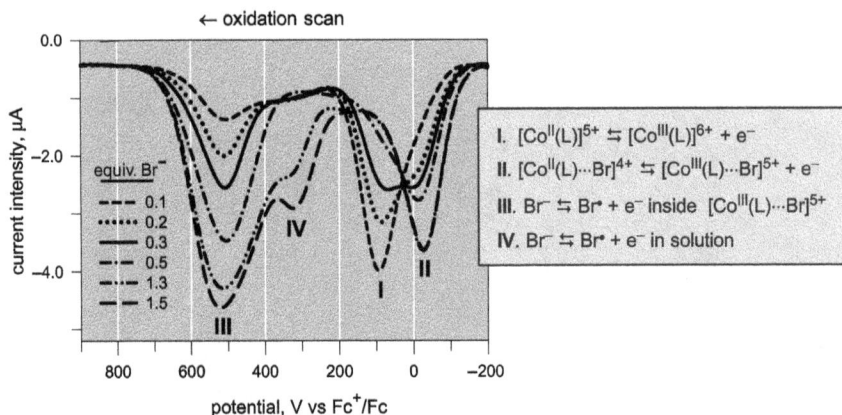

Figure 21.13. Family of DPV profiles taken at a platinum electrode over the course of the titration with [Bu$_4$N]Br of an MeCN solution of [Co^{II}(L)]$^{5+}$ (L = **55**) (potential E (mV) vs the Fc$^+$/Fc reference couple), potential scanning from negative to positive potentials (10 mV s^{-1}).[178]

Before bromide addition, peak **I** is observed, pertinent to the Co^{II}-to-Co^{III} oxidation of the empty pseudocryptand. Then following bromide addition and inclusion, peak **II** develops at a less positive potential, as observed in the titration with chloride. ΔE is in this case 120 mV, i.e. less than observed for chloride (ΔE = 144 mV), which indicates a less pronounced stabilisation of Co^{III} following anion inclusion (K^{III}/K^{II} = 107). Scanning has been in this case extended to 900 mV vs Fc$^+$/Fc, which shows the occurrence of further oxidation processes: **III**, the oxidation of the included Br$^-$ to the radical Br$^{\bullet}$, which takes place during the addition of the first equivalent of Br$^-$; then, **IV**, on addition of the second equivalent, the uncomplexed Br$^-$ undergoes easier oxidation at a less positive potential. The problem is that in the case of more reducing anions, e.g. I$^-$ and NCS$^-$, the oxidation peaks of the complexed and uncomplexed anion overlap the peaks pertinent to the oxidation of the metal centre, preventing any electrochemical characterisation.

It was then considered that, in order to separate the peaks pertinent to the [Co^{II}(L)]$^{5+}$/[Co^{III}(L)]$^{6+}$ redox process from those originating from the oxidation of the anion, one should carry out investigations in a medium of higher polarity. In fact, a more polar medium is expected to stabilise electrically charged species with respect to species which are neutral or detain a lower electrical charge. Therefore, in a more polar medium, (i) X^- will be stabilised over $X^\bullet$, thus making $E(X^\bullet/X^-)$ increase, and (ii) $[Co^{III}(L)]^{6+}$ will be stabilised over $[Co^{II}(L)]^{5+}$, thus making $E(Co^{III}/Co^{II})$ decrease. As a consequence, potentials pertinent to the oxidation of the metal centre and to the anion could be conveniently separated. Therefore, voltammetric titrations were performed in MeCN/H$_2$O solution 4:1 v/v. Rather surprisingly, the electrochemical response was quite different from that observed in pure MeCN. In particular, on anion addition, the DPV peak associated to the $[Co^{II}(L)]^{5+}/[Co^{III}(L)]^{6+}$ oxidation process did not decrease in intensity (and no new peak was observed at a less positive potential), but it simply shifted towards more negative potentials. Figure 21.14(a) shows the family of voltammetric profiles obtained on titration with chloride.

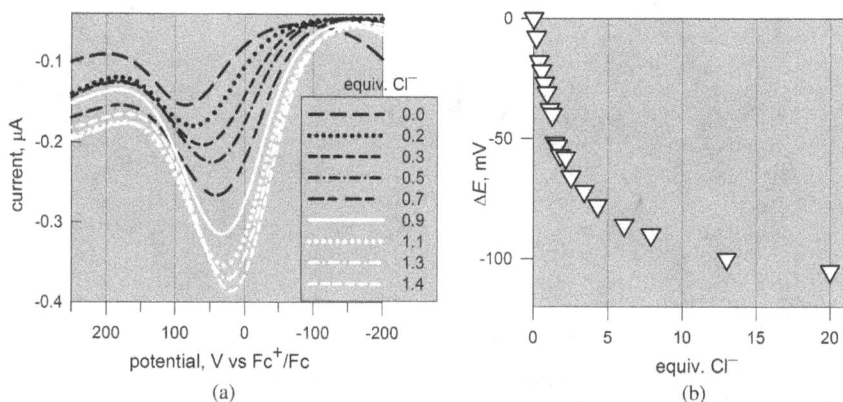

Figure 21.14. (a) DPV profiles obtained at a platinum working electrode over the course of the titration of a MeCN/H$_2$O solution (4:1 v/v) 7×10^{-4} M of $[Co^{II}(L)]^{5+}$ (L = **55**) and 0.05 M in [Bu$_4$N]PF$_6$ with a standard solution of [BnBu$_3$N]Cl. Oxidation scan from -200 mV to 200 mV vs Fc$^+$/Fc (10 mV s^{-1}). (b) Decrease of ΔE on addition of chloride; ΔE = potential measured during addition of Cl$^-$ — potential in the absence of Cl$^-$.[178]

In particular, the oxidation peak, on anion addition, is shifted towards less positive potentials, according to a saturation profile, illustrated in Figure 21.14(b). Such a behaviour (continuous shift of the potential) contrasts with that observed for the system and illustrated in Figure 21.12 (rise and fall of two peaks, each positioned at a fixed potential). This behaviour can be explained by assuming that the pseudocryptand $[Co^{III}(L)]^{5+}$ in aqueous MeCN does include an anion X^-, whereas $[Co^{II}(L)]^{4+}$ does not. This is due to the high energy cost associated with the dehydration of the anion. Thus, the redox behaviour of the $[Co^{III}(L)]^{6+}/[Co^{II}(L)]^{5+}$ in the presence of X^- must be described by the triangular scheme illustrated in Figure 21.15(b), from which it derives that ΔE linearly depends upon $[X^-]$. Notice that in the case of chloride, as well as for the other investigated anions (*vide infra*), the linear dependence of ΔE upon $[X^-]$ is not observed. This may be due to the fact that K^{II} has a small, but not negligible value.

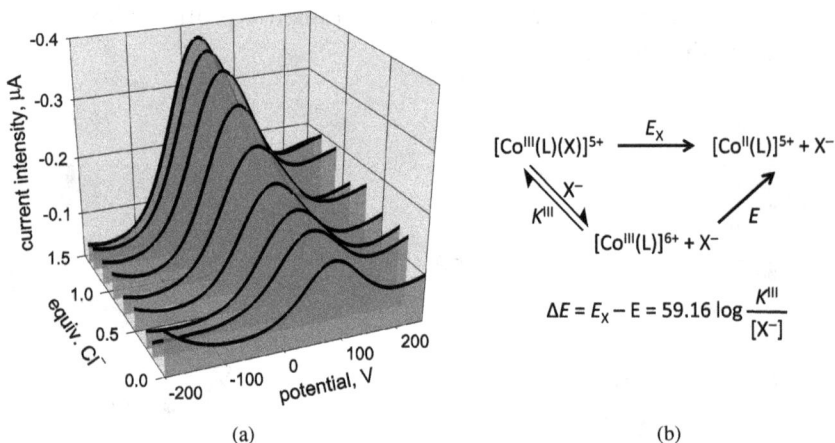

$$[Co^{III}(L)(X)]^{5+} \xrightarrow{E_X} [Co^{II}(L)]^{5+} + X^-$$

$$[Co^{III}(L)]^{6+} + X^-$$

$$\Delta E = E_X - E = 59.16 \log \frac{K^{III}}{[X^-]}$$

(a)   (b)

Figure 21.15. (a) DPV profiles of Figure 21.14a drawn in a 3D mode; (b) thermodynamic cycle for the $[Co^{II}(L)]^{5+}/[Co^{III}(L)]^{6+}$ (L = **55**) redox change in the presence of the anion X^-, which forms a stable 1:1 complex only with the receptor in the highest oxidation state, $[Co^{III}(L)]^{6+}$. K^{III} is the constant of the association equilibrium of the receptor $[Co^{III}(L)]^{6+}$ with X^-. $[Co^{III}(L)]^{5+}$ does not form a stable inclusion complex with X^-.[178]

Noticeably, in the MeCN/H_2O medium, other halide and pseudohalide anions shows a well-defined separation between the peaks associated with the Co^{III}/Co^{II} redox change and the peaks related to anion oxidation. In all cases, a continuous shift of the Co^{III}/Co^{II} potential was observed on anion addition. Pertinent profiles of the Co^{III}/Co^{II} redox couple potential are shown in Figure 21.16. Also in the present case, the dependence of ΔE upon $[X^-]$ is not linear, which reflects an intermediate situation between the square scheme in Figure 21.12(b) and the triangular scheme in Figure 21.15(b).

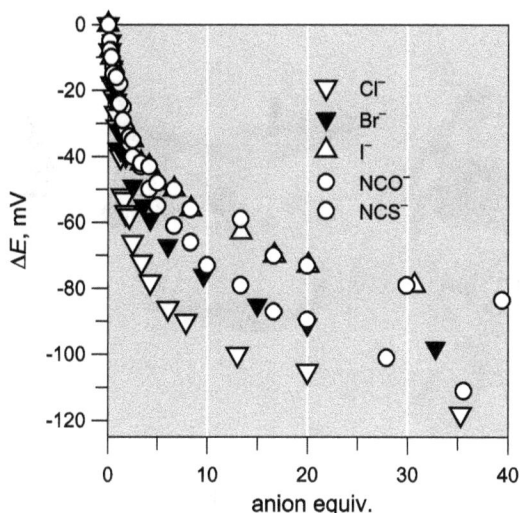

Figure 21.16. Titration of a MeCN/water solution (4:1 v/v) of $[Co^{II}(L)]^{5+}$ with halide and pseudohalide tetraalkylammonium salts. ΔE expresses the shift of the potential of the Co^{III}/Co^{II} couple induced by excess anion addition. L = **55**.

ΔE values at the highest investigated anion excess increase along the series: $I^- \sim NCS^- < NCO^- \sim Br^- < Cl^-$. This sequence parallels the anion tendency to receive H-bonds from the acidic C–H fragments of the pseudocryptand and confirms the major role played by hydrogen bonding interactions in the stabilisation of the $[Co^{III}(L)(X)]^{5+}$ complex over the $[Co^{II}(L)(X)]^{4+}$ complex.

A further pseudocryptand of the family of tripodand **55** was obtained by using as a locking metal Ru^{II}.[179] The $Ru^{II}(bpy)_3{}^{2+}$ subunit guarantees irreversible closure of the cavity and a luminescent signal, which could be, hopefully, modulated by anion inclusion. Crystals were obtained of the complex salt $[Ru^{II}(L)\cdots N_3](PF_6)_4 \cdot 3H_2O$ (L = **55**), whose structure is shown in Figure 21.17(a).[180] Not surprisingly, the coordination mode of azide in the $[Ru^{II}(L)]^{5+}$ pseudocryptand is similar to that observed for the $[Fe^{II}(L)]^{5+}$ analogue: $N_3{}^-$ is partially inserted into the receptor's cavity, with its axis normal to the line joining the centroid of the 1,3,5-trimethylbenzene cap and ruthenium(II). However, the anion penetration into the pseudocryptand is distinctly less pronounced than for Fe^{II} (see Figures 21.17(c) and 21.17(d)). This does not prevent the establishing

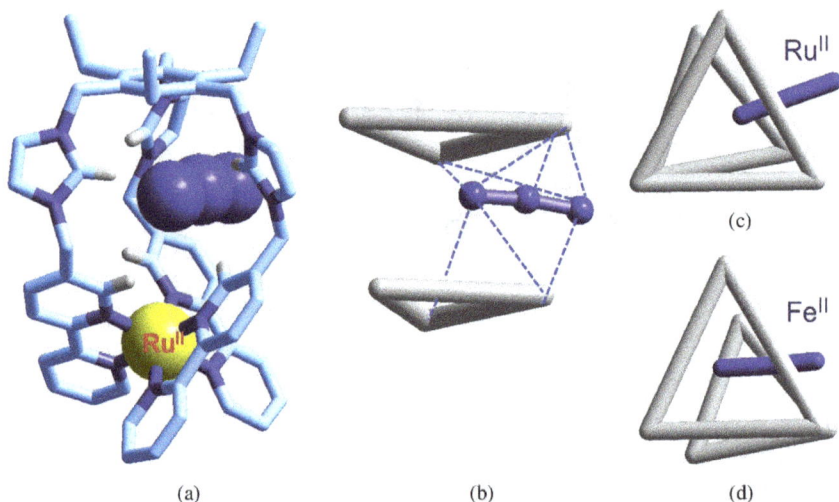

(a)   (b)   (c)

(d)

Figure 21.17. (a) the crystal structure of the complex salt $[Ru^{II}(L)\cdots N_3](PF_6)_4\cdot 3H_2O$ (L = **55**);[180] all hydrogens but H(im) and H(py) atoms, hexafluorophosphate counterions and solvate water molecules have been omitted for clarity; (b) nearly equilateral triangles obtained by linking the three H(im) (upper) and the three H(py) (lower); blue dashed lines indicate hydrogen bonds of low energy, to which C–H$\cdots$N distances ranging from 2.30 to 2.95 Å correspond; (c) and (d): top view of the triangles for pseudocryptates of Ru^{II} and Fe^{II}.

of a dense network of H-bonds, which involves all the nitrogen atoms of N_3^-, for a total of eight (Figure 21.17(b)).

The capability of a given system to act as a luminescent sensor is typically tested by carrying out a spectrofluorimetric titration. In the present case, for comparative purposes, an MeCN/H$_2$O solution (4:1 v/v) of the reference system $[Ru^{II}(bpy)_3]^{2+}$ was titrated with $[Bu_4]I$.[179] Iodide was chosen in view of its marked reducing properties and its involvement in electron transfer processes. In particular, I$^-$ addition caused a moderate decrease of the luminescent emission. In Figure 21.18(a), the ratio of the emission intensity at 600 nm before anion addition, I_0, over the emission intensity during anion addition, I, is plotted vs the molar concentration of I$^-$. The linear dependence indicates that the kinetics of the process obeys to the Stern–Volmer relationship. Thus, luminescence quenching has a *dynamic* nature and takes place through the occasional collisions between I$^-$ and $[Ru^{II}(bpy)_3]^{2+}$.

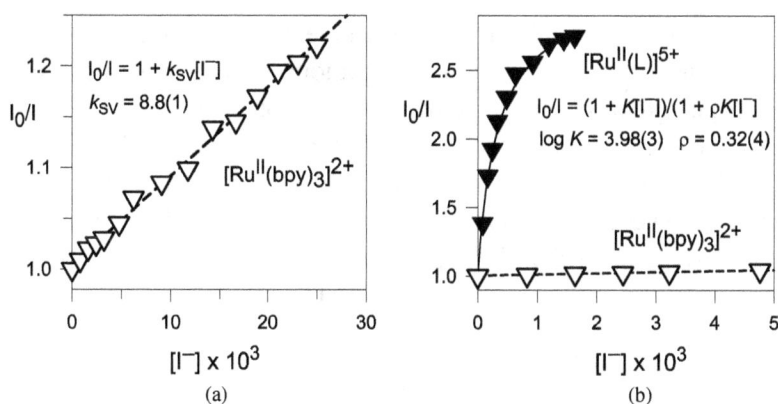

Figure 21.18. Titration with [Bu$_4$N]I of an MeCN/H$_2$O solution (4:1 v/v) of (a) [Ru^{II}(bpy)$_3$]$^{2+}$ and (b) of [Ru^{II}(L)]$^{5+}$ (L = **55**): k_{SV} is the Stern–Volmer constant, K is the formation constant of the [Ru^{II}(L)···I]$^{4+}$ pseudocryptate, ρ is the residual luminescence emitted by [Ru^{II}(L)···I]$^{4+}$ in the presence of an excess of I$^-$. Quenching of [Ru^{II}(bpy)$_3$]$^{2+}$ obeys the Stern–Volmer relationship, indicating a *dynamic* quenching mechanism; in the case of [Ru^{II}(L)]$^{5+}$ the I_0/I plot reveals a *static* mechanism of quenching.

Then, the same titration experiment was carried out on a solution of [Ru^{II}(L)]$^{5+}$ (L = **55**). Iodide addition induced a substantial quenching of the luminescent emission at 605 nm, while the I_0/I vs [I$^-$] plot showed a saturation behaviour, indicating a *static* quenching mechanism (see Figure 21.18(b)). In particular, data fitted the equation illustrated in figure, which describes the following mechanistic behaviour: (i) [Ru^{II}(L)]$^{5+}$ and I$^-$ form the [Ru^{II}(L)···I]$^{4+}$ inclusion complex, characterised by the equilibrium constant K; (ii) the encapsulated iodide transfers an electron to the proximate *Ru^{II}(bpy)$_3$$^{2+}$ subunit, quenching its emission to 42% (residual fluorescence ρ) (iii) occasional collisions by the I$^-$ ions added in excess do not quench the residual fluorescence, indicating that the pseudocryptand superstructure exerts in some way an efficient shield to the occurrence of the intermolecular electron transfer. Notice that in Figure 21.18(b), the I_0/I vs [I$^-$] plot of [Ru^{II}(bpy)$_3$]$^{2+}$ has been reported in the same scale used for the pseudocryptate complex. Comparison demonstrates that the designed positioning of iodide close to the luminophore has a quenching effect significantly higher with respect to statistical collisions.

These studies have illustrated the efficiency and versatility of metal-locked pseudocryptands in anion recognition: choice of the metal determines the nature of sensing, electrochemical or photochemical. Successful examples described here refer to systems formally outside of the declared limits of this book (bicyclic systems based on two $N(CH_2CH_2)_3$ capping subunits). However, their mention can be useful in stimulating and inspiring the developments that this topic deserves.

22

Use of Cryptands to Make Automatic Molecular Burettes

There exists a tendency in the field of Chemistry to create molecular devices, i.e. low molecular weight systems capable of performing a desired function inspired from everyday practice.[181] Thus, a variety of molecules or supramolecular systems of varying sophistication have been synthesised during the last two decades, which behave as (i) machines,[182] linear[183] and rotary[184] motors, brakes,[185] gears,[186] ratchets[187] (mechanical inspiration), and (ii) batteries,[188] electrical wires,[189] extendable cables,[190] plug–socket connectors,[191] switches[192] (electrical inspiration). Some other molecules are addressed to hi-tech functions from electronics and computer science and behave as logic gates,[193] adders and subtractors,[194] calculators,[195] and memory storage devices.[196] The design of these systems cannot be considered as a mere academic exercise or a proud exhibition of the power of chemical synthesis. Noticeably, the Nobel Prize in Chemistry 2016 was awarded jointly to Jean-Pierre Sauvage, Fraser Stoddart and Ben Feringa *'for the design and synthesis of molecular machines'*. In particular, as Olof Ramström, Member of the Nobel

Committee for Chemistry, stated: 'we are at the dawn of a new industrial revolution of the 21st century, and the future will show how molecular machinery can become an integral part of our lives. The advances made have also led to the first steps towards creating truly programmable machines, and it can be envisaged that *molecular robotics* will be one of the next major scientific areas'.[197]

Ironically, a few molecular devices have been inspired from chemical practice. One of these is the *molecular burette* invented by Giuseppe Alibrandi, Università di Messina, Italy, capable of performing pH-itrations from the inside of the solution.[198] The burette is based on the unique acid–base behaviour of the 1.1.1-crypt, **56**, as illustrated in Figure 22.1.[199,200]

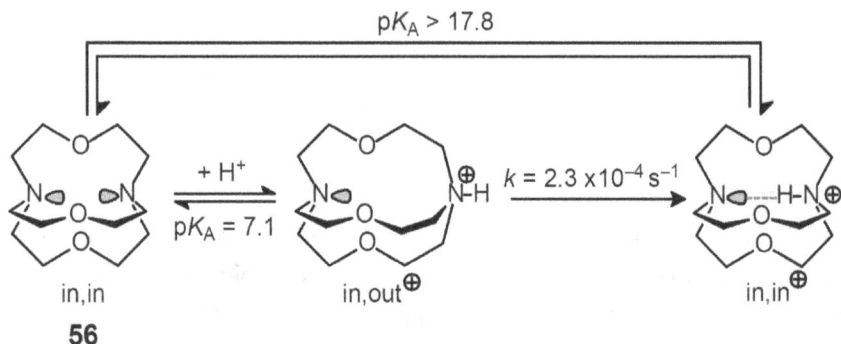

Figure 22.1. The unique acid–base properties of the 1.1,1-crypt (**56**, L): $L_{in,in}$ is a weak base (pK_A of the conjugate acid $LH^+_{in,out}$ = 7.1); $LH^+_{in,out}$ undergoes a slow conformational rearrangement (τ = 72.5 min) to the extremely stable entrapped proton species $LH^+_{in,in}$, the conjugate acid of a very strong base (p$K_A \geq$ 17.8). Thus, L, dissolved in water, slowly releases OH⁻, thereby increasing pH.[199,200]

The 1.1.1-crypt (L) is present in a solution as the (*in,in*) conformer. $L_{in,in}$ is a weak base (pK_A = 7.1) and, in a 10^{-3} M solution, only 1% of it reacts with H_2O to give $LH^+_{in,out}$. Then, the $LH^+_{in,out}$ conformer undergoes a slow rearrangement (τ = 72.5 min) to $LH^+_{in,in}$, an extremely stable species containing an entrapped proton. Stability is achieved from the presence of a rather strong hydrogen bond between the N–H fragment pointing inside and the lone pair of the facing tertiary amine nitrogen atom, still pointing inside, as illustrated in Figure 22.1. The high stability of $LH^+_{in,in}$ is reflected in the value of its pK_A (≥17.8). Thus, L, as the conjugate base of an extremely weak acid, is an extremely strong base (a proton sponge) and, when dissolved in water, produces protons slowly and increasingly, guaranteeing a linear increase in pH with a slope of 10^{-4} pH s^{-1}.

In the first experiment by Alibrandi and coworkers,[198] to a cuvette containing a solution of 10^{-2} M in 1.1.1-crypt, adjusted to pH = 6.5 with HBF$_4$, one drop of a concentrated solution of phenol red (ca. 10^{-4} M) was added. Then, pH slowly increased over the 6.5–8.5 interval, while the colour of the indicator changed from yellow to fuchsia, and the absorption band centred at λ = 558 nm (pertaining to the deprotonated phenol red species) smoothly increased. From the titration data, a value of pK_A = 7.90 ± 0.01 was determined for phenol red, compared to the literature value of 7.92.[201]

Moving to a more sophisticated apparatus, an automatic chloride dispenser has been designed by coupling the functions displayed by 1.1.1-crypt (**56**) and the dicopper(II) bistren cryptate including a Cl$^-$ ion, [Cu$^{II}_2$(L)(Cl)]$^{3+}$ (L = **31**).[202] Its mechanism is illustrated in Figure 22.2.

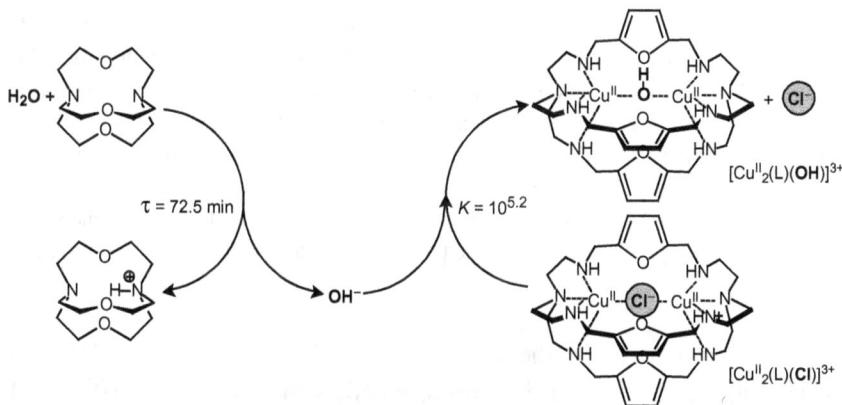

Figure 22.2. An automatic dispenser of chloride: the anion *container* is the complex [Cu$^{II}_2$(L)(Cl)]$^{3+}$ (L = **31**); the *pump* is 1.1.1-crypt, which slowly releases OH$^-$. The hydroxide ions slowly displace Cl$^-$ from [Cu$^{II}_2$(L)(Cl)]$^{3+}$; chloride is released to the solution at a programmed rate.[202]

The chloride *container* is the complex [Cu$^{II}_2$(L)(Cl)]$^{3+}$, the *pump* is 1.1.1-crypt. The small cryptand pumps OH$^-$ ions at a programmed rate, which replace Cl$^-$ ions in the container: chloride is released into the solution. Based on these propositions, a solution containing the following substances was prepared: 1.1.1-cryptand (0.1 M), L (= **31**, 0.003 M), $Cu(CF_3SO_3)_2$ (0.006 M), KCl (0.003 M) and triflic acid was added to adjust pH_{init} to 6. Then, the spectra of the solution were recorded every 15 minutes over a period of 6 hours. The correspondent family of spectra is reported in Figure 22.3(a).

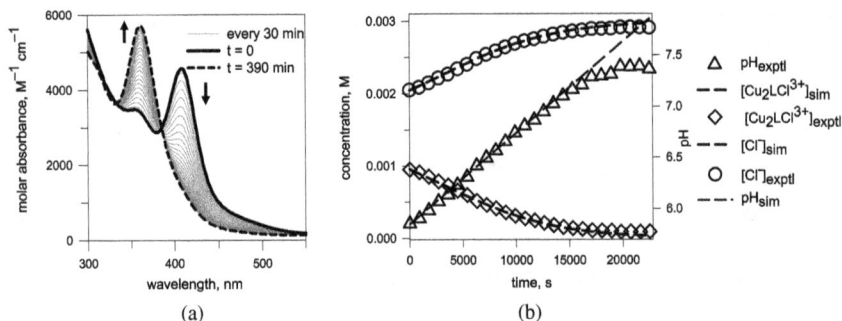

(a)   (b)

Figure 22.3.   (a) Family of spectra of a solution containing 1.1.1-crypt (0.1 M), L (= **31**, 0.003 M), $Cu(CF_3SO_3)_2$ (0.006 M), KCl (0.003 M) and adjusted to pH = 6 with triflic acid: the decreasing band at 410 nm pertains to the $[Cu^{II}_2(L)(Cl)]^{3+}$ complex (the container of Cl^-), the increasing band centred at 350 nm pertains to $[Cu^{II}_2(L)(OH)]^{3+}$ complex, spectral changes indicate the production of OH^- ions by the 1.1.1-crypt, which goes on to displace Cl^- ion from the container complex; (b) symbols: quantities obtained from the 'automatic' spectrophotometric titration, pH was measured with a glass electrode; lines are the simulated concentration profiles, pH is expected to increase at a rate of 1.0×10^{-4} unit s^{-1}.

Over the course of the automatic titration, the band at 410 nm ($[Cu^{II}_2(L)(Cl)]^{3+}$ complex) decreased, whereas that at 350 nm ($[Cu^{II}_2(L)(OH)]^{3+}$ complex) increased, monitoring the chloride release. For every OH^- ion generated by the pump (1.1.1-crypt), one Cl^- is released from the container to the solution. Figure 22.3(b) compares the experimental concentration profiles with those calculated with appropriate equations, showing a satisfactory agreement. However, the performance of the automatic dispenser is far from perfect. In fact, the pump cannot operate at a pH lower than 6.5, and, at this pH, the $[Cu^{II}_2(L)(Cl)]^{3+}$ complex is present only at 30% and $[Cu^{II}_2(L)(OH)]^{3+}$ complex at 70%. Thus, it happens that the dispenser, at the beginning of the automatic titration, has already lost 2/3 of its contents and can release to the solution only the remaining 1/3. In particular, as shown by the circles in Figure 22.3(b), at $t = 0$, the concentration of free Cl^- is already 2×10^{-3} M (circles), and at the end of the titration, it has been brought to 3×10^{-3} M. Indeed, the design of the dispenser is fine, but the pump will need to be replaced!

23

Cages in Everyday Life, Chemistry and Art

The tendency of chemists to give molecular systems names borrowed from the everyday life was discussed in Chapter 22. In this perspective, cryptands rightfully belong to the family of *cages*. Cages are familiar objects for humankind since at least ten thousands years. Humans build and use cages, but do not love them for several reasons: (i) they are a symbol of forced spatial constriction and freedom deprivation for living beings, (ii) they exhibit private details of the imprisoned individual to the public, affirming a state of weakness and dependence, (iii) the typical guest of cages is a tender and undefended being (a canary, a parrot). Thus, humans use cages mostly for leisure, but are not proud of this practise and are reluctant to its emphasis.

It is probably for this reason that artists have not been inspired by cages, in painting and handicraft. There are only two relevant paintings featuring a cage and they are quite recent on the time scale of art history. One has been painted by René Magritte (1898–1967) — *Elective Affinities*, 1933, shown in Figure 23.1. The prisoner of the cage is an egg, and probably the artist intended to bewilder the viewer by illustrating the delayed affinity of two objects to each other, the cage and the egg, from which the typical guest of the cage, a bird, originates. Someone could get a distressful message: every man, even before his birth, is destined to live within the narrow limits and the severe rules of a merciless society.

Figure 23.1. René Magritte, *Elective Affinities* (1933) — oil on canvas. Private collection.

In chemistry things are different: synthesising a molecular cage and confining in it a chemical species (a metal ion, an anion or a molecule) is considered a deserving and admirable action. The design and synthesis of cages at a molecular level has become a so popular and distinguished activity to be mentioned in the Merriam-Webster Dictionary: 'an arrangement of atoms or molecules so bonded as to enclose a space in which another atom or ion (as of a metal) can reside'.[203] However, to be more precise, the analogy between the cages of the macroscopic world and those of the molecular world may not be fully justified. In actuality, a living being in a macroscopic cage stays under a nasty *kinetic* control: it may have (and, in general, does have) a great tendency to escape from the cage, but this event is prevented by an insurmountable activation barrier (dense metal bars and a firmly locked gate). On the molecular side, such a kinetically controlled situation has been observed only in rare cases (e.g. in transition metal complexes of sarcophagines). On the other hand, no or a very moderate kinetic barrier between the inner and the outer state exists for *s* block metal ions (e.g. in the case of classical cryptands) and for anions (e.g. in the case of bistren cryptands and cryptates): the guest (no longer a prisoner) can get in or out at will and its sojourn in the cage is *thermodynamically* controlled.

The second example of cages in figurative arts is given by the famous and intriguing wood engraving print by M. C. Escher (*Stars*, 1948). The print depicts two chameleons confined in a cage composed of three interlocking regular octahedra, floating through space (see Figure 23.2(a)). The image seems to illustrate the attempt of the Universe to impose its immutable and celestial order (represented by the polyhedral shapes) to the overwhelming force of Life (the two chameleons).

 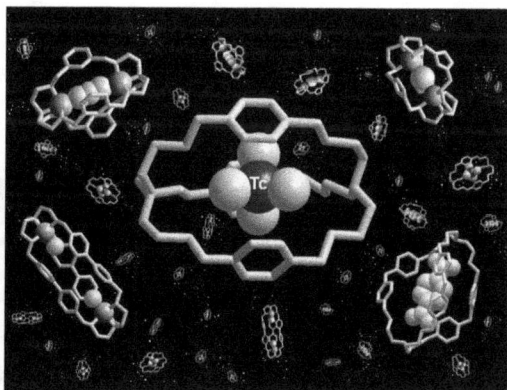

(a)   (b)

Figure 23.2. (a) M. C. Escher, *Stars*, 1948, wood engraved print; (b) *Organic and Biomolecular Chemistry*, front cover, vol. 13, number 12, 28 March 2015. Reproduced from Ref. 204 with permission from the Royal Society of Chemistry.

As most chemists, we were impressed by this picture and, when asked by the Editor of *Organic and Biomolecular Chemistry* to illustrate the cover of an issue in which was included a review paper by us on bistren cryptands,[204] we prepared a graphics inspired by the Escher's print, which is shown in Figure 23.2(b). The artistic quality of the picture is undoubtedly poor, but the drawn chemical cages remain fascinating and promising objects deserving attention and further investigations by the younger generations of chemists.

References

1. C. J. Pedersen, *J. Am. Chem. Soc.*, **1967**, *89*, 2495.
2. C. J. Pedersen, *J. Am. Chem. Soc.*, **1967**, *89*, 7017.
3. C. J. Pedersen, The discovery of crown ethers, Nobel Lecture (1987), https://www.nobelprize.org/nobel_prizes/chemistry/laureates/1987/pedersen-lecture.html.
4. J. A. A. de Boer, D. N. Reinhoudt, S. Harkema, G. J. van Hummel, F. de Jong, *J. Am. Chem. Soc.*, **1982**, *104*, 4073.
5. J. D. Dunitz, P. Seiler, *Acta Crystallogr., Sect.B: Struct. Crystallogr. Cryst. Chem.*, **1974**, *30*, 2739.
6. S. J. Donnie, H. E. Simmons, *J. Am. Chem. Soc.*, **1972**, *94.11*, 4024–4025.
7. H. Bock, D. Jaculi, *Angew. Chem. Int. Ed. Eng.*, **1984**, *23*, 305.
8. H. K. Frensdorff, *J. Am. Chem. Soc.*, **1971**, *93*, 600.
9. A. Alvanipour, J. L. Atwood, S. G. Bott, P. C. Junk, U. H. Kynast, H. Prinz, *J. Chem. Soc., Dalton Trans.*, **1998**, 1223.
10. P. Seiler, M. Dobler, J. D. Dunitz, *Acta Crystallogr., Sect.B: Struct. Crystallogr. Cryst. Chem.*, **1974**, *30*, 2744.
11. S. P. Petrosyants, A. B. Ilyukhin, *Koord. Khim. (Russ. Coord. Chem.)*, **2007**, *33*, 747.
12. M. Dobler, R. P. Phizackerley, *Acta Crystallogr., Sect.B: Struct. Crystallogr. Cryst. Chem.*, **1974**, *30*, 2748.

13. H. Bock, C. Näther, Z. Havlas, A. John, C. Arad, *Angew. Chem., Int. Ed. Eng.*, **1994**, *33*, 875.
14. L. Bonomo, E. Solari, R. Scopelliti, C. Floriani, *Chem. — Eur. J.*, **2001**, *7*, 1322.
15. M. K. Chantooni Jr., D. Britton, I. M. Kolthoff, *J. Crystallogr. Spectrosc. Res.*, **1993**, *23*, 497.
16. T. Mandai, S. Tsuzuki, K. Ueno, K. Dokko, M. Watanabe, *Phys. Chem. Chem. Phys. (PCCP)*, **2015**, *17*, 2838.
17. G. D. Smith, W. L. Duax, D. A. Langs, G. T. DeTitta, J. W. Edmonds, D. C. Rohrer, C. M. Weeks, *J. Am. Chem. Soc.*, **1975**, *97*, 7242.
18. J. A. Hamilton, M. N. Sabesan, L. K. Steinrauf, *J. Am. Chem. Soc.*, **1981**, *103*, 5880.
19. B. Dietrich, J.-M. Lehn, J.-P. Sauvage, *Tetrahedron Lett.*, **1969**, *10*, 2889.
20. B. Metz, D. Moras, R. Weiss, *J. Chem. Soc., Perkin Trans.* **1976**, *2*, 423.
21. D. Moras, B. Metz, R. Weiss, *Acta Crystallogr., Sect. B: Struct. Crystallogr. Cryst. Chem.*, **1973**, *29*, 383.
22. J.-M. Lehn, J.-P. Sauvage, *J. Am. Chem. Soc.*, **1975**, *97*, 6700.
23. H. M. N. H. Irving, R. J. P. Williams, *J. Chem. Soc.*, **1953**, 3192.
24. B. Dietrich, J. M. Lehn, J. P. Sauvage, *Tetrahedron*, **1973**, *29*, 1647.
25. B. Dietrich, J. M. Lehn, J. P. Sauvage, *Tetrahedron Lett.*, **1969**, *10*, 2885.
26. B. Dietrich, J. M. Lehn, J. P. Sauvage, J. Bianzat, *Tetrahedron*, **1973**, *29*, 1629.
27. L. Arnaudet, P. Charpin, G. Folcher, M. Lance, M. Nierlich, D. Vigner, *Organometallics*, **1986**, *5*, 270.
28. F. Ettel, G. Huttner, L. Zsolnai, C. Emmerich, *J. Organomet. Chem.*, **1991**, *414*, 71.
29. I. A. Guzei, L. C. Spencer, J. W. Su, R. R. Burnette, *Acta Crystallogr., Sect.B: Struct. Sci.*, **2007**, *63*, 93.
30. R. W. Schmid, C. N. Reilley, *Anal. Chem.*, **1957**, *29*, 264.
31. M. Cotrait, *Acta Crystallogr., Sect.B: Struct. Crystallogr. Cryst. Chem.*, **1972**, *28*, 781.
32. J. J. Stezowski, R. Countryman, J. L. Hoard, *Inorg. Chem.*, **1973**, *12*, 1749.
33. A. S. Antsyshkina, G. G. Sadikov, A. L. Poznyak, V. S. Sergienko, *Zh. Neorg. Khim. (Russ. J. Inorg. Chem.)*, **2002**, *47*, 43.
34. R. Y. Tsien, *Biochemistry*, **1980**, *19*, 2396.
35. G. Grynkiewicz, M. Poenie, R. Y. Tsien, *J. Biol. Chem.*, **1985**, *260*, 3440.
36. A. Nezu, A. Tanimura, T. Morita, Y. Tojyo, *J. Cell. Sci.*, **2010**, *123*, 2292.
37. K. N. Trueblood, C. B. Knobler, E. F. Maverick, R. C. Helgeson, S. B. Brown, D. J. Cram, *J. Am. Chem. Soc.*, **1981**, *103*, 5594.

38. W. A. Henderson, N. R. Brooks, W. W. Brennessel, V. G. Young Jr, *J. Phys. Chem. A*, **2004**, *108*, 225.
39. D. J. Cram, G. M. Lein, T. Kaneda, R. C. Helgeson, C. B. Knobler, E. Maverick, K. N. Trueblood, *J. Am. Chem. Soc.*, **1981**, *103*, 6228.
40. D. J. Cram, G. M. Lein, *J. Am. Chem. Soc.*, **1985**, *107*, 3657.
41. R. A. Marcus, *J. Chem. Phys.*, **1956**, *24*, 966.
42. D. J. Cram, R. A. Carmack, M. P. deGrandpre, G. M. Lein, I. Goldberg, C. B. Knobler, E. F. Maverick, K. N. Trueblood, *J. Am. Chem. Soc.*, **1987**, *109*, 7068.
43. K. N. Trueblood, E. F. Maverick, C. B. Knobler, I. Goldberg, *Acta Crystallogr., Sect.C: Cryst. Struct. Commun.*, **1995**, *51*, 894.
44. J. R. Moran, St. Karbach, D. J. Cram, *J. Am. Chem. Soc.*, **1982**, *104*, 5826.
45. S. P. Artz, D. J. Cram, *J. Am. Chem. Soc.*, **1984**, *106*, 2160.
46. A. D. Bokare, A. Patnaik, *Cryst. Res. Technol.*, **2004**, *39*, 465.
47. R. M. Izatt, J. S. Bradshaw, S. A. Nielsen, J. D. Lamb, J. J. Christensen, D. Sen, *Chem. Rev.*, **1985**, *85*, 271.
48. A. N. Chekhlov, *Zh. Strukt. Khim. (J. Struct. Chem.)*, **2002**, *43*, 949.
49. D. J. Cram, S. Peng-Ho, *J. Am. Chem. Soc.*, **1986**, *108*, 2998.
50. K. E. Koenig, G. M. Lein, P. Stuckler, T. Kaneda, D. J. Cram, *J. Am. Chem. Soc.*, **1979**, *101*, 3553.
51. E. Graf, J.-P. Kintzinger, J.-M. Lehn, J. LeMoigne, *J. Am. Chem. Soc.*, **1982**, *104*, 1672.
52. B. Metz, J. M. Rosalky, R. Weiss, *J. Chem. Soc., Chem. Commun.*, **1976**, 533.
53. A. L. Spek, G. Roelofsen, J. G. Noltes, A. H. Alberts, *Cryst. Struct. Commun.*, **1982**, *11*, 1863.
54. M. M. Zhao, Z. R. Qu, *Acta Crystallogr., Sect.C: Cryst. Struct. Commun.*, **2010**, *66*, m215.
55. C. B. Shoemaker, L. V. McAfee, D. P. Shoemaker, C. W. DeKock, *Acta Crystallogr., Sect.C: Cryst. Struct. Commun.*, **1986**, *42*, 1310.
56. A. N. Chekhlov, *Zh. Strukt. Khim. (J. Struct. Chem.)*, **2004**, *45*, 1136.
57. J. M. Lehn, *Supramolecular Chemistry: Concepts and Perspectives. A Personal Account*, VCH, 1995.
58. A. Werner, *Z. Anorg. Allg. Chem.*, **1893**, *3*, 267.
59. J. D. van der Waals, *Over de Continuïteit van den Gas en Vloeistoftoestand (On the continuity of the gas and liquid state)*, PhD Thesis, University of Leiden, 1873.
60. W. M. Latimer, W. H. Rodebush, *J. Am. Chem. Soc.*, **1920**, *42*, 1419.
61. C. H. Park, H. E. Simmons, *J. Am. Chem. Soc.*, **1968**, *90*, 2431.

62. R. A. Bell, G. G. Christoph, F. R. Fronczek, R. E. Marsh, *Science*, **1975**, *190*, 151–152.
63. J. W. Steed, D. R. Turner, K. Wallace, *Core Concepts in Supramolecular Chemistry and Nanochemistry*, John Wiley & Sons, 2007.
64. E. Graf, J.-M. Lehn, *J. Am. Chem. Soc.*, **1976**, *98*, 6403.
65. F. P. Schmidtchen, *Angew. Chem. Int. Ed. Engl.*, **1977**, *16*, 720.
66. J.-M. Lehn, S. H. Pine, E. Watanabe, A. K. Willard, *J. Am. Chem. Soc.*, **1977**, *99*, 6766.
67. E. J. Laskowski, D. M. Duggan, D. N. Hendrickson, *Inorg. Chem.*, **1975**, *14*, 2449.
68. R. T. Stibrany, J. A. Potenza, CSD Communication (2007).
69. S. S. Massoud, F. A. Mautner, M. Abu-Youssef, N. M. Shuaib, *Polyhedron*, **1999**, *18*, 2287.
70. A. Marzotto, D. A. Clemente, G. Valle, *Acta Crystallogr., Sect. C: Cryst. Struct. Commun.*, **1993**, *49*, 1252.
71. R. J. Motekaitis, A. E. Martell, J. M. Lehn, E. I. Watanabe, *Inorg. Chem.*, **1982**, *21*, 4253.
72. R. J. Motekaitis, A. E. Martell, B. Dietrich, J.-M. Lehn, *Inorg. Chem.*, **1984**, *23*, 1588.
73. J.-M. Lehn, E. Sonveaux, A. K. Willard, *J. Am. Chem. Soc.*, **1978**, *100*, 4914.
74. R. J. Motekaitis, A. E. Martell, I. Murase, J.-M. Lehn, M. W. Hosseini, *Inorg. Chem.*, **1988**, *27*, 3630.
75. B. Dietrich, J. Guilhem, J.-M. Lehn, C. Pascard, E. Sonveaux, *Helv. Chim. Acta*, **1984**, *67*, 91.
76. J. Jazwinski, J.-M. Lehn, D. Lilienbaum, R. Ziessel, J. Guilhem, C. Pascard, *J. Chem. Soc., Chem. Comm.*, **1987**, 1691.
77. D. McDowell, J. Nelson, *Tetrahedron Lett.*, **1988**, *29*, 385.
78. R. J. Motekaitis, P. R. Rudolf, A. E. Martell, A. Clearfield, *Inorg.Chem.*, **1989**, *28*, 112.
79. R. Menif, J. Reibenspies, A. E. Martell, *Inorg. Chem.*, **1991**, *30*, 3446.
80. A. D. Bond, S. Derossi, F. Jensen, F. B. Larsen, C. J. McKenzie, J. Nelson, *Inorg.Chem.*, **2005**, *44*, 5987.
81. L. Yang, Y. Li, X. Zhuang, L. Jiang, J. Chen, R. L. Luck, T. Lu, *Chem. — Eur. J.*, **2009**, *15*, 12399.
82. M. Śmiechowski, J. Stangret, *J. Phys. Chem. A*, **2007**, *111*, 2889.
83. P. Paoletti, M. Ciampolini, *Ric. Sci.*, **1963**, *3*, 399.
84. L. Fabbrizzi, P. Pallavicini, A. Perotti, L. Parodi, A. Taglietti, *Inorg. Chim. Acta*, **1995**, *238*, 5.

85. C. J. Harding, F. E. Mabbs, E. J. L. MacInnes, V. McKee, J. Nelson, *J. Chem. Soc., Dalton Trans.*, **1996**, 3227.

86. F. A. Mautner, C. N. Landry, A. A. Gallo, S. S. Massoud, *J. Mol. Struct.*, **2007**, *837*, 72.

87. S. Derossi, A. D. Bond, C. J. McKenzie, J. Nelson, *Acta Crystallogr., Sect.E: Struct. Rep. Online*, **2005**, *61*, m1379.

88. V. Amendola, G. Bergamaschi, M. Boiocchi, L. Fabbrizzi, A. Poggi, M. Zema, *Inorg. Chim. Acta*, **2008**, *361*, 4038.

89. V. Amendola, E. Bastianello, L. Fabbrizzi, C. Mangano, P. Pallavicini, A. Perotti, A. Manotti Lanfredi, F. Ugozzoli, *Angew. Chem., Int. Ed.*, **2000**, *39*, 2917.

90. C. J. Harding, V. McKee, J. Nelson, Q. Lu, *J. Chem. Soc., Chem. Commun.*, **1993**, 1768.

91. A. W. Czarnik, *Acc. Chem. Res.*, **1994**, *27*, 302.

92. A. P. de Silva, T. S. Moody, G. D. Wright, *Analyst*, **2009**, 134, 2385.

93. L. Fabbrizzi, M. Licchelli, A. Taglietti, *Dalton Trans.*, **2003**, 3471.

94. L. Fabbrizzi, I. Faravelli, G. Francese, M. Licchelli, A. Perotti, A. Taglietti, *Chem. Commun.*, **1998**, 971.

95. A, Metzger, E. V. Anslyn, *Angew. Chem., Int. Ed.*, **1998**, *37*, 649.

96. L. Fabbrizzi, N. Marcotte, F. Stomeo, A. Taglietti, *Angew. Chem., Int. Ed.*, **2002**, *41*, 3811.

97. L. Fabbrizzi, A. Leone, A. Taglietti, *Angew. Chem., Int. Ed.*, **2001**, *40*, 3066.

98. S. L. Wiskur, H. Ait-Haddou, J. J. Lavigne, E. V. Anslyn, *Acc. Chem. Res.*, **2001**, *34*, 963.

99. D. G. Nicholls, S. J. Ferguson, S. J. *Bioenergetics 3*, Academic Press: San Diego, CA, 2002.

100. J. T. Hancock, *Cell Signalling*, Longman, Harlow, UK, 1997.

101. D. E. Metzler, C. M. Metzler, *Biochemistry: The Chemical Reactions of Living Cells*, Academic Press, San Diego, 2003.

102. S. K. Kim, D. H. Lee, J.-I. Hong, J.-I. Yoon, *Acc. Chem. Res.*, **2009**, *42*, 23.

103. V. Amendola, G. Bergamaschi, A. Buttafava, L. Fabbrizzi, E. Monzani, *J. Am. Chem. Soc.*, **2010**, *132*, 147.

104. T. Yajima, G. Maccarrone, M. Takani, A. Contino, G. Arena, R. Takamido, M Hanaki, Y. Funahashi, A. Odani, O. Yamauchi, *Chem. — Eur. J.* **2003**, *9*, 3341.

105. K. N. Trueblood, P. Horn, V. Luzzati, *Acta Crystallogr.* **1961**, *14*, 965.

106. E. Shefter, K. N. Trueblood, *Acta Crystallogr.*, **1965**, *18*, 1067.

107. V. McKee, J. Nelson, R. M. Town, *Chem. Soc. Rev.*, **2003**, *32*, 309.

108. F. Arnaud-Neu, S. Fuangswasdi, B. Maubert, J. Nelson, V. McKee, *Inorg. Chem.*, **2000**, *39*, 573.

109. Y. Li, L. Jiang, X.-L. Feng, T,-B. Lu, *Cryst.Growth Des.*, **2006**, *6*, 1074.

110. M. A. Hossain, J. M. Llinares, S. Mason, P. Morehouse, D. Powell, K. Bowman-James, *Angew. Chem., Int. Ed.*, **2002**, *41*, 2335.

111. M. A. Hossain, P. Morehouse, D. Powell, K. Bowman-James, *Inorg. Chem.*, **2005**, *44*, 2143.

112. G. A. Jeffrey, W. Saenger, *Hydrogen Bonding in Biological Structures*, Springer-Verlag, Berlin, 1994.

113. M. J. Hynes, B. Maubert, V. McKee, R. M. Town, J. Nelson, *J. Chem. Soc., Dalton Trans.*, **2000**, 2853.

114. B. M. Maubert, J. Nelson, V. McKee, R. M. Town, I. Pal, *J. Chem. Soc., Dalton Trans.*, **2001**, 1395.

115. S. Mason, T. Clifford, L. Seib, K. Kuczera, K. Bowman-James, *J. Am. Chem. Soc.*, **1998**, *120*, 8899.

116. K. Yoshihara, T. Omori (Eds), *Topics in Current Chemistry: Technetium and Rhenium*, vol. 176, Springer-Verlag, Berlin Heidelberg, 1996.

117. V. Amendola, G. Alberti, G. Bergamaschi, R. Biesuz, M. Boiocchi, S. Ferrito, F.-P. Schmidtchen, *Eur. J. Inorg. Chem.*, **2012**, *21*, 3410.

118. R. O'Brien, J. E. Ladbury, B. Z. Chowdry, *Isothermal Titration Calorimetry of Biomolecules*, in *Protein–Ligand Interactions: Hydrodynamics and Calorimetry*, Eds: S. E. Harding, B. Z. Chowdry, Oxford University Press, 2000.

119. F.-P. Schmidtchen, *Isothermal titration calorimetry in supramolecular chemistry*, in *Supramolecular Chemistry: From Molecules to Nanomaterials*, Eds: P. Gale, J. Steed, John Wiley and Sons, 2014.

120. D. Farrell, K. Gloe, K. Gloe, G. Goretzki, V. McKee, J. Nelson, M. Nieuwenhuyzen, I. Pal, H. Stephan, R. M. Town, K. Wichmann, *Dalton Trans.*, 2003, 1961.

121. R. Alberto, G. Bergamaschi, H. Braband, T. Fox, V. Amendola, *Angew. Chem., Int. Ed.*, **2012**, *51*, 9772.

122. V. Amendola, G. Bergamaschi, M. Boiocchi, R. Alberto, H. Braband, *Chemical Science*, **2014**, *5*, 1820.

123. D. L. Dexter, *J. Chem. Phys.*, **1951**, *21*, *836*.

124. N. W. Alcock, K. Bowman-James, C. Miller, J. M. L. Berenguer, CSD Communication, 2005.

125. J.-M. Lehn, R. Meric, J.-P. Vigneron, I. Bkouche-Waksman, C. Pascard, *Chem. Commun.*, **1991**, 62.

126. P. Mateus, R. Delgado, V. André, M. T. Duarte, *Inorg. Chem.*, **2015**, *54*, 229.

127. I. A. Teslya, A. I. Tursina, L. D. Iskhakova, L. M. Avdonina, V. V. Marugin, *Zh. Strukt. Khim.*, **1990**, *31*, 123.

128. M. Boiocchi, M. Bonizzoni, L. Fabbrizzi, G. Piovani, A. Taglietti, *Angew. Chem., Int. Ed.*, **2004**, *43*, 3847.

129. W. J. McEntee, T. H. Crook, *Psychopharmacology*, **1993**, *111*, 391.

130. J. S. Marvin, B. G. Borghuis, L. Tian, J. Cichon, M. T. Harnett, J. Akerboom, A. Gordus, S. L. Renninger, T.-W. Chen, C. I. Bargmann, M. B. Orger, E. R. Schreiter, J. B. Demb, W.-B. Gan, S. A. Hires, L. L. Looger, *Nat. Meth.*, **2013**, *10*, 162.

131. M. Bonizzoni, L. Fabbrizzi, G. Piovani, A. Taglietti, *Tetrahedron*, **2004**, *60*, 11159.

132. S. Valiyaveettil, J. F. J. Engbersen, W. Verboom, D. N. Reinhoudt, *Angew. Chem., Int. Ed. Engl.*, **1993**, *32*, 900.

133. S. K. Dey, G. Das, *Chem. Commun.*, **2011**, *47*, 4983.

134. T. Steiner, *Angew. Chem. Int. Ed.*, **2002**, *41*, 48.

135. S. Saha, B. Akhuli, I. Ravikumar, P. S. Lakshminarayanan, P. Ghosh, *Cryst. Eng. Comm.*, **2014**, *16*, 4796.

136. S. O. Kang, J. M. Llinares, V. W. Day, K. Bowman-James, *Chem. Soc. Rev.*, **2010**, *39*, 3980.

137. S. O. Kang, J. M. Llinares, D. Powell, D. VanderVelde, Kristin Bowman-James, *J. Am. Chem. Soc.*, **2003**, *125*, 10152.

138. S. O. Kang, D. VanderVelde, D. Powell, K. Bowman-James, *J. Am. Chem. Soc.*, **2004**, *126*, 12272.

139. S. O. Kang, D. Powell, V. W. Day, K. Bowman-James, *Angew. Chem., Int. Ed.*, **2006**, *45*, 1921.

140. W. Gauss, P. Moser, G. Schwarzenbach, *Helv. Chim. Acta*, **1952**, *35*, 2359.

141. S. O. Kang, V. W. Day, K. Bowman-James, *Inorg. Chem.*, **2010**, *49*, 8629.

142. J. A. Ibers, *J. Chem. Phys.*, **1964**, *40*, 402.

143. M. Boiocchi, L. Del Boca, D. Esteban Gómez, L. Fabbrizzi, M. Licchelli, E. Monzani, *J. Am. Chem. Soc.*, **2004**, *126*, 16507.

144. M. Boiocchi, L. Del Boca, D. Esteban Gómez, L. Fabbrizzi, M. Licchelli, E. Monzani, *Chem. — Eur. J.*, **2005**, *11*, 3097.

145. D. Esteban-Gómez, L. Fabbrizzi, M. Licchelli, E. Monzani, *Org. Biomol. Chem.*, **2005**, *3*, 1495.

146. J. Emsley, *Chem. Soc. Rev.*, **1980**, *9*, 91.

147. V. Amendola, D. Esteban-Gómez, L. Fabbrizzi, M. Licchelli, *Acc. Chem. Res.*, **2006**, *39*, 343.

148. L. Vaska, J. W. DiLuzio, *J. Am. Chem. Soc.*, **1961**, *83*, 2784.

149. N. Lopez, D. J. Graham, R. McGuire Jr, G. E. Alliger, Y. Shao-Horn, C. C. Cummins, D. G. Nocera, *Science*, **2012**, *335*, 450.

150. B. L. Anderson, A. G. Maher, M. Nava, N. Lopez, C. C. Cummins, D. G. Nocera, *J. Phys. Chem. B*, **2015**, *119*, 7422.

151. Y. Sheng, I. A. Abreu, D. E. Cabelli, M. J. Maroney, A.-F. Miller, M. Teixeira, J. S. Valentine, *Chem. Rev.*, **2014**, *114*, 3854.

152. H. J. Lawson, J. D. Atwood, *J. Am. Chem. Soc.*, **1989**, *111*, 6223.

153. M. Nava, N. Lopez, P. Muller, G. Wu, D. G. Nocera, C. C. Cummins, *J. Am. Chem. Soc.*, **2015**, *137*, 14562.

154. P. H. Smith, M. E. Barr, J. R. Brainard, D. K. Ford, H. Freiser, S. Muralidharan, S. D. Reilly, R. R. Ryan, L. A. Silks, W.-H. Yu, *J. Org. Chem.*, **1993**, *58*, 7939.

155. I. I. Creaser, J. M. Harrowfield, A. J. Herlt, A. M. Sargeson, J. Springborg, R. J. Geue, M. R. Snow, *J. Am. Chem. Soc.*, **1977**, *99*, 3181.

156. R. J. Geue, T. W. Hambley, J. M. Harrowfield, A. M. Sargeson, M. R. Snow, *J. Am. Chem. Soc.*, **1984**, *106*, 5478.

157. G. A. Bottomley, I. J. Clark, I. I. Creaser, L. M. Engelhardt, R. J. Geue, K. S. Hagen, J. M. Harrowfield, G. A. Lawrance, P. A. Lay, A. M. Sargeson, A. J. See, B. W. Skelton, A. H. White, F. R. Wilner, *Aust. J. Chem.*, **1994**, *47*, 143.

158. I. I. Creaser, L. M. Engelhardt, J. M. Harrowfield, A. M. Sargeson, B. W. Skelton, A. H. White, *Aust. J. Chem.*, **1993**, *46*, 465.

159. P. Comba, I. I. Creaser, L. R. Gahan, J. M. Harrowfield, G. A. Lawrance, L. L. Martin, A. W. H. Mau, A. M. Sargeson, W. H. F. Sasse, M. R. Snow, *Inorg. Chem.*, **1986**, *25*, 384.

160. I. J. Clark, A. Crispini, P. S. Donnelly, L. M. Engelhardt, J. M. Harrowfield, S.-H. Jeong, Y. Kim, G. A. Koutsantonis, Y. H. Lee, N. A. Lengkeek, M. Mocerino, G. L. Nealon, M. I. Ogden, Y. C. Park, C. Pettinari, L. Polanzan, E. Rukmini, A. M. Sargeson, B. W. Skelton, A. N. Sobolev, P. Thuery, A. H. White, *Aust. J. Chem.*, **2009**, *62*, 1246.

161. J. A. Thompson, M. E. Barr, D. K. Ford, L. A. Silks III, J. McCormick, P. H. Smith, *Inorg. Chem.*, **1996**, *35*, 2025.

162. G. De Santis, L. Fabbrizzi, A. Perotti, N. Sardone, A. Taglietti, *Inorg. Chem.*, **1997**, *36*, 1998.

163. J. Coyle, A. J. Downard, J. Nelson, V. McKee, C. J. Harding, R. Herbst-Irmer, *Dalton Trans.*, **2004**, 2357.

164. D. Farrell, C. J. Harding, V. McKee, J. Nelson, *Dalton Trans.*, **2006**, 3204.

165. M. Paik Suh, J. W. Jeon, H. R. Moon, K. S. Min, H. J. Choi, *C. R. Chim.*, **2005**, *8*, 1543.

166. M. Schatz, M. Becker, O. Walter, G. Liehr, S. Schindler, *Inorg. Chim. Acta*, **2001**, *324*, 173.

167. N. Zheng, X. Bu, P. Feng, *Chem. Commun.*, **2005**, 2805.

168. M. E. Barr, P. H. Smith, W. E. Antholine, B. Spencer, *Chem. Commun.*, **1993**, 1649.

169. B. Dietrich, B. Dilworth, J.-M. Lehn, J. Souchez, M. Cesario, J. Guilhem, C. Pascard, *Helv. Chim. Acta*, **1996**, *79*, 569.

170. M. Arunachalam, E. Suresh, P. Ghosh, *Tetrahedron*, **2007**, *63*, 11371.

171. M. A. Hossain, J. M. Llinares, C. A. Miller, L. Seib, K. Bowman-James, *Chem. Commun.*, **2000**, 2269.

172. B. Zhang, P. Cai, C. Duan, R. Miao, L. Zhu, T. Niitsu, H. Inoue, *Chem. Commun.*, **2004**, 2206.

173. V. Amendola, L. Fabbrizzi, C. Mangano, A. M. Lanfredi, P. Pallavicini, A. Perotti, F. Ugozzoli, *J. Chem. Soc., Dalton Trans.*, **2000**, 1155.

174. N. C. A. Baker, N. C. Fletcher, P. N. Horton, M. B. Hursthouse, *Dalton Trans.*, **2012**, *41*, 7005.

175. S. Campagna, F. Puntoriero, F. Nastasi, G. Bergamini, V. Balzani, *Top. Curr. Chem.*, **207**, *280*, 117.

176. R. D. Shannon, *Acta Cryst*, **1976**, *A32*, 751.

177. V. Amendola, M. Boiocchi, B. Colasson, L. Fabbrizzi, M.-J. Rodriguez-Douton, F. Ugozzoli, *Angew. Chem., Int. Ed.*, **2006**, *45*, 6920.

178. V. Amendola, M. Boiocchi, B. Colasson, L. Fabbrizzi, E. Monzani, M.-J. Douton-Rodriguez, C. Spadini, *Inorg. Chem.*, **2008**, *47*, 4808.

179. V. Amendola, B. Colasson, L. Fabbrizzi, unpublished results.

180. V. Amendola, M. Boiocchi, B. Colasson, L. Fabbrizzi, CSD Communication.

181. V. Balzani, A. Credi, M. Venturi, *Molecular Devices and Machines — Concepts and Perspective for the Nanoworld*, Wiley-VCH, Weinheim, 2008.

182. A. Coskun, M. Banaszak, R. D. Astumian, J. F. Stoddart, B. A. Grzybowski, *Chem. Soc. Rev.*, **2012**, *41*, 19.

183. J.-P. Collin, C. Dietrich-Buchecker, P. Gaviña, M. C. Jimenez-Molero, J.-P. Sauvage, *Acc. Chem. Res.*, **2001**, *34*, 477.

184. N. Ruangsupapichat, M. M. Pollard, S. R. Harutyunyan, B. L. Feringa, *Nat. Chem.*, **2011**, *3*, 53.

185. K. Nikitin, C. Bothe, H. Müller-Bunz, Y. Ortin, M. J. McGlinchey, *Organometallics*, **2012**, *31*, 6183.

186. D. K. Frantz, A. Linden, K. K. Baldridge, J. S. Siegel, *J. Am. Chem. Soc.*, **2012**, *134*, 1528.

187. M. Alvarez-Pérez, S. M. Goldup, D. A. Leigh, A. M. Z. Slawin, *J. Am. Chem. Soc.*, **2008**, *130*, 1836.

188. C. Floriani, E. Solari, F. Franceschi, R. Scopelliti, P. Belanzoni, M. Rosi, *Chem. — Eur. J.*, **2001**, *7*, 3052.

189. G. J. E. Davidson, S. J. Loeb, P. Passaniti, S. Silvi, A. Credi, *Chem. — Eur. J.*, **2006**, *12*, 3233.

190. R. Ballardini, V. Balzani, M. Clemente-León, A. Credi, M. T. Gandolfi, E. Ishow, J. Perkins, J. F. Stoddart, H.-R. Tseng, S. Wenger, *J. Am. Chem. Soc.*, **2002**, *124*, 12786.

191. G. Rogez, B. Ferrer-Ribera, A. Credi, R. Ballardini, M. T. Gandolfi, V. Balzani, Y. Liu, B. H. Northrop, J. F. Stoddart, *J. Am. Chem. Soc.*, **2007**, *129*, 4633.

192. B. L. Feringa, W. R. Browne (Eds), *Molecular Switches*, Wiley, New York, 2011.

193. A. P. DeSilva, *Chem. Asian J.*, **2011**, *6*, 750.

194. D. Margulies, G. Melman, A. Shanzer, *J. Am. Chem. Soc.*, **2006**, *128*, 4865.

195. D. Margulies, G. Melman, A. Shanzer, *Nat. Mater.*, **2005**, *4*, 768.

196. U. Pischel, *Angew. Chem., Int. Ed.*, **2010**, *49*, 1356.

197. *Scientific Background on the Nobel Prize in Chemistry 2016*, The Royal Academy of Science, 5 October 2016.

198. G. Alibrandi, C. Lo Vecchio, G. Lando, *Angew. Chem., Int. Ed.*, **2009**, *48*, 6332.

199. J. Cheney, J.-M. Lehn, *J. Chem. Soc., Chem. Commun.*, **1972**, 487.

200. P. B. Smith, J. L. Dye, J. Cheney, J. M. Lehn, *J. Am. Chem. Soc.*, **1981**, *103*, 6044.

201. *Handbook of Analytical Chemistry*, Ed. L. Meites, McGraw-Hill, New York, 1963, Sec. 3, p. 36.

202. G. Alibrandi, V. Amendola, G. Bergamaschi, R. Dollenz, L. Fabbrizzi, M. Licchelli, C. Lo Vecchio, *Chem. — Eur. J.*, **2013**, *19*, 3729.

203. See https://www.merriam-webster.com/dictionary/cage, retrieved on 9 September 2017.

204. A. P. Davis, S. Kubik, A. Dalla Cort (Guest Eds), Themed issue on 'Supramolecular chemistry in water', *Org. Biomol. Chem.*, **2015**, *13*, 2499.

Index